Soil Science and Management

Soil Science and Management

Edited by
Leyton Gray

Larsen & Keller
www.larsen-keller.com

Soil Science and Management
Edited by Leyton Gray
ISBN: 978-1-63549-263-7 (Hardback)

Larsen & Keller
Published by Larsen and Keller Education,
5 Penn Plaza,
19th Floor,
New York, NY 10001, USA

Cataloging-in-Publication Data

Soil science and management / edited by Leyton Gray.
p. cm.
Includes bibliographical references and index.
ISBN 978-1-63549-263-7
1. Soil science. 2. Soil management. 3. Soils. 4. Soil pollution. I. Gray, Leyton.
S591 .S65 2017
631.4--dc23

Printed and bound in China.

For more information regarding Larsen and Keller Education and its products, please visit the publisher's website www.larsen-keller.com

Table of Contents

Preface

Soil management refers to the practices, techniques and methods used in order to protect soil and improve its quality. It is performed to restore and maintain soil fertility and to increase crop yield. In this text, chapters are complied in such a manner that they provide detailed information about the different soil managing techniques. It attempts to understand the multiple branches that fall under the discipline of soil management and how such concepts have practical applications. For someone with an interest and eye for detail, this text covers the most significant topics of this subject.

To facilitate a deeper understanding of the contents of this book a short introduction of every chapter is written below:

Chapter 1- Soil management concerns all the operations, practices and treatments used to protect soil and it enhances the performance of soil. It is an integral part of land management, and focuses on the different types of soil and their characteristics. The chapter on soil management offers an insightful focus, keeping in mind the complex subject matter.

Chapter 2- Soil is a fundamental part of the nature and the natural environment. It is equally as important as plants, animals, rocks, landforms and rivers. Soil is a mixture of minerals, organic matter, gases, and the countless organisms that together support life on earth. The following chapter inquires profoundly into the topic of soil.

Chapter 3- Some of the properties discussed in this chapter are the texture of soil, soil structure, peds, soil color, soil resistivity and soil water. Soil structure is the arrangement of the parts of soil and the pore space located between them. Soil color tells us what is specific about the soil, for instance, soil high in iron is deep orange brown to yellowish brown while the soil that has high content of organic matter is dark brown or black. This chapter is an overview of the subject matter incorporating all the major aspects of soil properties.

Chapter 4- The study of soil as a natural resource is soil science. Geologists, physical geographers, ecologists, microbiologists and specialists in regional planning all contribute to the further knowledge of soil and the advancement of the soil sciences. The essential concepts of soil science discussed in this chapter are soil compaction, soil health, soil pH, nitrogen fixation and tilth. The chapter strategically encompasses and incorporates the major components and key concepts of soil science, providing a complete understanding.

Chapter 5- The harm caused by the presence of human made chemicals and industrial waste in the soil is soil pollution. The cause of this pollution is industrial activities, agricultural chemicals and improper disposal of waste. The following content explains to the reader soil retrogression and degradation.

Chapter 6- Techniques are important components of any field of study. The following chapter elucidates the various fundamental techniques and approaches of soil management. It serves as a source to understand soil conservation, liming, soil conditioner, cover crop and nutrient management. This chapter discusses the methods of soil management in a critical manner providing key analysis to the subject matter.

I would like to share the credit of this book with my editorial team who worked tirelessly on this book. I owe the completion of this book to the never-ending support of my family, who supported me throughout the project.

Editor

1

Introduction to Soil Management

Soil management concerns all the operations, practices and treatments used to protect soil and it enhances the performance of soil. It is an integral part of land management, and focuses on the different types of soil and their characteristics. The chapter on soil management offers an insightful focus, keeping in mind the complex subject matter.

Environment

According to the EPA, agricultural soil management practices can lead to production and emission of nitrous oxide (N2O), a major greenhouse gas and air pollutant. Activities that can contribute to N2O emissions include fertilizer usage, irrigation and tillage. The management of soils accounts for over half of the emissions from the Agriculture sector. Cattle livestocks account for one third of emissions, through methane emissions. Manure management and rice cultivation also emit emissions.

Methods that significantly enhance carbon sequestration in soil include no-till farming, residue mulching, cover cropping, and crop rotation, all of which are more widely used in organic farming than in conventional farming. Because only 5% of US farmland currently uses no-till and residue mulching, there is a large potential for carbon sequestration.

Practices

Soil management practices that affect soil quality:

- Controlling traffic on the soil surface helps to reduce soil compaction, which can reduce aeration and water infiltration.
- Cover crops keep the soil anchored and covered in off-seasons so that the soil is not eroded by wind and rain.
- Crop rotations for row crops alternate high-residue crops with lower-residue crops to increase the amount of plant material left on the surface of the soil during the year to protect the soil from erosion.
- Nutrient management can help to improve the fertility of the soil and the amount of organic matter content, which improves soil structure and function.

- Tillage, especially reduced-tillage or no-till operations limit the amount of soil disturbance while cultivating a new crop, and help to maintain plant residues on the surface of the soil for erosion protection and water retention.

Advantages of Soil Management

- Maintain soil fertility
- Restore soil fertility
- Make the agricultural process an economic one
- Help increase yield

2

A Brief Introduction to Soil

Soil is a fundamental part of the nature and the natural environment. It is equally as important as plants, animals, rocks, landforms and rivers. Soil is a mixture of minerals, organic matter, gases, and the countless organisms that together support life on earth. The following chapter inquires profoundly into the topic of soil.

Soil is a mixture of minerals, organic matter, gases, liquids, and countless organisms that together support life on Earth. Soil is a natural body called the pedosphere which has four important functions: it is a medium for plant growth; it is a means of water storage, supply and purification; it is a modifier of Earth's atmosphere; it is a habitat for organisms; all of which, in turn, modify the soil.

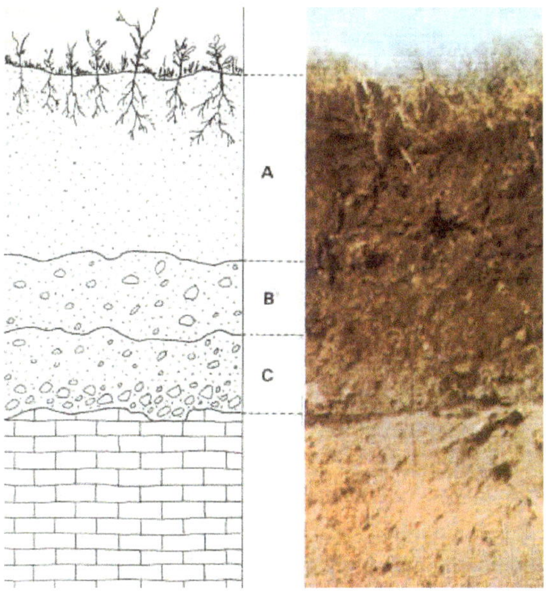

A represents soil; B represents laterite, a regolith; C represents saprolite, a less-weathered regolith; the bottom-most layer represents bedrock.

Soil is called the "Skin of the Earth" and interfaces with its lithosphere, hydrosphere, atmosphere, and biosphere. The term pedolith, used commonly to refer to the soil, literally translates 'level stone'. Soil consists of a solid phase of minerals and organic matter, as well as a porous phase that holds gases and water. Accordingly, soils are often treated as a three-state system of solids, liquids, and gases.

Soil is a product of the influence of the climate, relief (elevation, orientation, and slope of terrain), organisms, and its parent materials (original minerals) interacting over time. Soil continually undergoes development by way of numerous physical, chemical and biological processes, which include weathering with associated erosion.

Surface-water-gley developed in glacial till, Northern Ireland.

Most soils have a density between 1 and 2 g/cm³. Little of the soil of planet Earth is older than the Pleistocene and none is older than the Cenozoic, although fossilized soils are preserved from as far back as the Archean.

Soil science has two basic branches of study: edaphology and pedology. Edaphology is concerned with the influence of soils on living things. Pedology is focused on the formation, description (morphology), and classification of soils in their natural environment. In engineering terms, soil is referred to as regolith, or loose rock material that lies above the 'solid geology'. Soil is commonly referred to as "earth" or "dirt"; technically, the term "dirt" should be restricted to displaced soil.

As soil resources serve as a basis for food security, the international community advocates its sustainable and responsible use through different types of Soil Governance.

Overview

Soil is a major component of the Earth's ecosystem. The world's ecosystems are impacted in far-reaching ways by the processes carried out in the soil, from ozone depletion and global warming, to rainforest destruction and water pollution. Following the atmosphere, the soil is the next largest carbon reservoir on Earth, and it is potentially one of the most reactive to human disturbance and climate change. As the planet warms, soils will add carbon dioxide to the atmosphere due to its increased biological activity at higher temperatures. Thus, soil carbon losses likely have a large positive feedback response to global warming.

Soil acts as an engineering medium, a habitat for soil organisms, a recycling system for nutrients and organic wastes, a regulator of water quality, a modifier of atmospheric composition, and a medium for plant growth. Since soil has a tremendous range of available niches and habitats, it contains most of the Earth's genetic diversity. A gram of soil can contain billions of organisms, belonging to thousands of species. Soil has a mean prokaryotic density of roughly 10^{13} organisms per cubic meter, whereas the ocean has a mean prokaryotic density of roughly 10^8 organisms per cubic meter. The carbon content of the soil is eventually returned to the atmosphere through the process of respiration carried out by heterotrophic organisms that feed upon the carbonaceous material in the soil. Since plant roots need oxygen, ventilation is an important characteristic of soil. This ventilation can be accomplished via networks of soil pores, which also absorb and hold rainwater making it readily available for plant uptake. Since plants require a nearly continuous supply of water, but most regions receive sporadic rainfall, the water-holding capacity of soils is vital for plant survival.

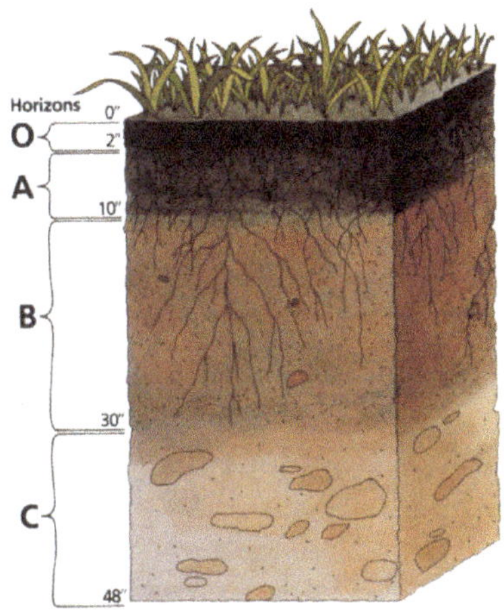

Soil Profile: Darkened topsoil and reddish subsoil layers are typical in some regions.

Soils can effectively remove impurities, kill disease agents, and degrade contaminants. Typically, soils maintain a net absorption of oxygen and methane, and undergo a net release of carbon dioxide and nitrous oxide. Soils offer plants physical support, air, water, temperature moderation, nutrients, and protection from toxins. Soils provide readily available nutrients to plants and animals by converting dead organic matter into various nutrient forms.

Soils supply plants with nutrients that are held in place by the clay and humus content of that soil. For optimum plant growth, the soil components by volume should be roughly 50% solids (45% mineral and 5% organic matter), and 50% voids of which half is occupied by water and half by gas. The percent soil mineral and organic content is typically treated as a constant, while the percent soil water and gas content is considered highly variable whereby a rise in one is simultaneously balanced by a reduction in the other. The pore space allows for the infiltration and movement of air and water, both of which are critical for life in soil. Compaction, a common problem with soils, reduces this space, preventing air and water from reaching the plant roots and soil organisms.

Given sufficient time, an undifferientated soil will evolve a soil profile which consists of two or more layers, referred to as soil horizons, that differ in one or more properties such as in their texture, structure, density, porosity, consistency, temperature, color, and reactivity. The horizons differ greatly in thickness and generally lack sharp boundaries. Soil profile development is dependent on the processes that form soils from their parent materials, the type of parent material, and the factors that control soil formation. The biological influences on soil properties are strongest near the surface, while the geochemical influences on soil properties increase with depth. Mature soil profiles in temperate climate regions typically include three basic master horizons: A, B and C. The solum normally includes the A and B horizons. The living component of the soil is largely confined to the solum. In the more hot, humid, climate of the tropics, a soil may have only a single horizon.

The soil texture is determined by the relative proportions of sand, silt, and clay in the soil. The addition of organic matter, water, gases and time causes the soil of that texture to develop into a larger soil structure called an aggregate. At that point a soil can be said to be developed, and can be described further in terms of color, porosity, consistency, reaction etc.

Of all the factors influencing the evolution of soil, water is the most powerful due to its involvement in the solution, erosion, transportation, and deposition of the materials of which a soil is composed. The mixture of water and the dissolved or suspended materials that occupy the soil pore space is called the soil solution. Since soil water is never pure water, but contains hundreds of dissolved organic and mineral substances, it may be more accurately called the soil solution. Water is central to the solution, precipitation and leaching of minerals from the soil profile. Finally, water affects the type of vegetation that grows in a soil, which in turn affects the development of the soil.

The most influential factor in stabilizing soil fertility are the soil colloids, clay and humus. Soil colloids behave as repositories of nutrients and moisture and so act to buffer the variations of soil solution ions and moisture. The contribution of soil colloids to soil nutrition are out of proportion to their part of the soil. Colloids act to store nutrients that might otherwise be leached from the soil or to release those ions in response to changes of soil pH, and so, make them available to plants.

The greatest influence on plant nutrient availability is soil pH, which is a measure of the hydrogen ion (acid-forming) soil reactivity, and is in turn a function of the soil materials, precipitation level, and plant root behavior. Soil pH strongly affects the availability of nutrients.

Most nutrients, with the exception of nitrogen, originate from minerals. Some nitrogen originates from rain (as dilute nitric acid), but most of the nitrogen is available in soils as a result of nitrogen fixation by bacteria. The action of microbes on organic matter and minerals may be to free nutrients for use, sequester them, or cause their loss from the soil by their volatilisation to gases or their leaching from the soil. The nutrients may be stored on soil colloids, or within live or dead organic matter, but they may not be accessible to plants due to extremes of pH.

The organic material of the soil has a powerful effect on its development, fertility, and available moisture. Following water and soil colloids, organic material is next in importance to a soil's formation and fertility.

History of The Study of Soil

Studies Concerning Soil Fertility

The history of the study of soil is intimately tied to our urgent need to provide food for ourselves and forage for our animals. Throughout history, civilizations have prospered or declined as a function of the availability and productivity of their soils.

The Greek historian Xenophon (450–355 B.C.) is credited with being the first to expound upon the merits of green-manuring crops: "But then whatever weeds are upon the ground, being turned into earth, enrich the soil as much as dung."

Columella's "Husbandry," circa 60 A.D., advocated the use of lime and that clover and alfalfa (green manure) should be turned under, and was used by 15 generations (450 years) under the Roman Empire until its collapse. From the fall of Rome to the French Revolution, knowledge of soil and agriculture was passed on from parent to child and as a result, crop yields were low. During the European Dark Ages, Yahya Ibn al-'Awwam's handbook, with its emphasis on irrigation, guided the people of North Africa, Spain and the Middle East; a translation of this work was finally carried to the southwest of the United States.

Experiments into what made plants grow first led to the idea that the ash left behind when plant matter was burned was the essential element but overlooked the role of nitrogen, which is not left on the ground after combustion. In about 1635, the Flemish chemist Jan Baptist van Helmont thought he had proved water to be the essential element from his famous five years' experiment with a willow tree grown with only the addition of rainwater. His conclusion came from the fact that the increase in the plant's weight had apparently been produced only by the addition of water, with no reduction in the soil's weight. John Woodward (d. 1728) experimented with various types of water ranging from clean to muddy and found muddy water the best, and so he concluded that earthy matter was the essential element. Others concluded it was humus in the soil that passed some essence to the growing plant. Still others held that the vital growth principal was something passed from dead plants or animals to the new plants. At the start of the 18th century, Jethro Tull demonstrated that it was beneficial to cultivate (stir) the soil, but his opinion that the stirring made the fine parts of soil available for plant absorption was erroneous.

As chemistry developed, it was applied to the investigation of soil fertility. The French chemist Antoine Lavoisier showed in about 1778 that plants and animals must "combust" oxygen internally to live and was able to deduce that most of the 165-pound weight of van Helmont's willow tree derived from air. It was the French agriculturalist Jean-Baptiste Boussingault who by means of experimentation obtained evidence showing that the main sources of carbon, hydrogen and oxygen for plants were the air and water. Justus von Liebig in his book *Organic Chemistry in its Applications to Agriculture and Physiology* (published 1840), asserted that the chemicals in plants must have come from the soil and air and that to maintain soil fertility, the used minerals must be replaced. Liebig nevertheless believed the nitrogen was supplied from the air. The enrichment of soil with guano by the Incas was rediscovered in 1802, by Alexander von Humboldt. This led to its mining and that of Chilean nitrate and to its application to soil in the United States and Europe after 1840.

The work of Liebig was a revolution for agriculture, and so other investigators started experimentation based on it. In England John Bennet Lawes and Joseph Henry Gilbert worked in the Rothamsted Experimental Station, founded by the former, and discovered that plants took nitrogen from the soil, and that salts needed to be in an available state to be absorbed by plants. Their investigations also produced the "superphosphate", consisting in the acid treatment of phosphate rock. This led to the invention and use of salts of potassium (K) and nitrogen (N) as fertilizers. Ammonia generated by the production of coke was recovered and used as fertiliser. Finally, the chemical basis of nutrients delivered to the soil in manure was understood and in the mid-19th century chemical fertilisers were applied. However, the dynamic interaction of soil and its life forms awaited discovery.

In 1856 J. T. Way discovered that ammonia contained in fertilisers was transformed into nitrates, and twenty years later R. W. Warington proved that this transformation was done by living organisms. In 1890 Sergei Winogradsky announced he had found the bacteria responsible for this transformation.

It was known that certain legumes could take up nitrogen from the air and fix it to the soil but it took the development of bacteriology towards the end of the 19th century to lead to an understanding of the role played in nitrogen fixation by bacteria. The symbiosis of bacteria and leguminous roots, and the fixation of nitrogen by the bacteria, were simultaneously discovered by German agronomist Hermann Hellriegel and Dutch microbiologist Martinus Beijerinck.

Crop rotation, mechanisation, chemical and natural fertilisers led to a doubling of wheat yields in Western Europe between 1800 and 1900.

Studies Concerning Soil Formation

The scientists who studied the soil in connection with agricultural practices had considered it mainly as a static substrate. However, soil is the result of evolution from more ancient geological materials. After studies of the improvement of the soil commenced, others began to study soil genesis and as a result also soil types and classifications.

In 1860, in Mississippi, Eugene W. Hilgard studied the relationship among rock material, climate, and vegetation, and the type of soils that were developed. He realised that the soils were dynamic, and considered soil types classification. Unfortunately his work was not continued. At the same time Vasily Dokuchaev (about 1870) was leading a team of soil scientists in Russia who conducted an extensive survey of soils, finding that similar basic rocks, climate and vegetation types lead to similar soil layering and types, and established the concepts for soil classifications. Due to the language barriers, the work of this team was not communicated to Western Europe until 1914 by a publication in German by K. D. Glinka, a member of the Russian team.

Curtis F. Marbut was influenced by the work of the Russian team, translated Glinka's publication into English, and as he was placed in charge of the U. S. National Cooperative Soil Survey, applied it to a national soil classification system.

Soil-Forming Processes

Soil formation, or pedogenesis, is the combined effect of physical, chemical, biological and anthropogenic processes working on soil parent material. Soil is said to be formed when organic matter has accumulated and colloids are washed downward, leaving deposits of clay, humus, iron oxide, carbonate, and gypsum, producing a distinct layer called the B horizon. This is a somewhat arbitrary definition as mixtures of sand, silt, clay and humus will support biological and agricultural activity before that time. These constituents are moved from one level to another by water and animal activity. As a result, layers (horizons) form in the soil profile. The alteration and movement of materials within a soil causes the formation of distinctive soil horizons.

How soil formation proceeds is influenced by at least five classic factors that are intertwined in the evolution of a soil. They are: parent material, climate, topography (relief), organisms, and time. When reordered to climate, relief, organisms, parent material, and time, they form the acronym CROPT.

An example of the development of a soil would begin with the weathering of lava flow bedrock, which would produce the purely mineral-based parent material from which the soil texture forms. Soil development would proceed most rapidly from bare rock of recent flows in a warm climate, under heavy and frequent rainfall. Under such conditions, plants become established very quickly on basaltic lava, even though there is very little organic material. The plants are supported by the porous rock as it is filled with nutrient-bearing water that carries dissolved minerals from the rocks and guano. Crevasses and pockets, local topography of the rocks, would hold fine materials and harbour plant roots. The developing plant roots are associated with mycorrhizal fungi that assist in breaking up the porous lava, and by these means organic matter and a finer mineral soil accumulate with time.

Parent Material

The mineral material from which a soil forms is called parent material. Rock, whether its origin is igneous, sedimentary, or metamorphic, is the source of all soil mineral materials and the origin of all plant nutrients with the exceptions of nitrogen, hydrogen and carbon. As the parent material is chemically and physically weathered, transported, deposited and precipitated, it is transformed into a soil.

Typical soil parent mineral materials are:

- Quartz: SiO_2
- Calcite: $CaCO_3$
- Feldspar: $KAlSi_3O_8$
- Mica (biotite): $K(Mg,Fe)_3AlSi_3O_{10}(OH)_2$

Classification of Parent Material

Parent materials are classified according to how they came to be deposited. Residual materials are mineral materials that have weathered in place from primary bedrock. Transported materials are those that have been deposited by water, wind, ice or gravity. Cumulose material is organic matter that has grown and accumulates in place.

Loess field in Germany.

Residual soils are soils that develop from their underlying parent rocks and have the same general chemistry as those rocks. The soils found on mesas, plateaux, and plains are residual soils. In the United States as little as three percent of the soils are residual.

Most soils derive from transported materials that have been moved many miles by wind, water, ice and gravity.

- Aeolian processes (movement by wind) are capable of moving silt and fine sand many hundreds of miles, forming loess soils (60–90 percent silt), common in the Midwest of North America and in Central Asia. Clay is seldom moved by wind as it forms stable aggregates.
- Water-transported materials are classed as either alluvial, lacustrine, or marine. Alluvial materials are those moved and deposited by flowing water. Sedimentary deposits settled in lakes are called lacustrine. Lake Bonneville and many soils around the Great Lakes of the United States are examples. Marine deposits, such as soils along the Atlantic and Gulf Coasts and in the Imperial Valley of California of the United States, are the beds of ancient seas that have been revealed as the land uplifted.
- Ice moves parent material and makes deposits in the form of terminal and lateral moraines in the case of stationary glaciers. Retreating glaciers leave smoother ground moraines and in all cases, outwash plains are left as alluvial deposits are moved downstream from the glacier.
- Parent material moved by gravity is obvious at the base of steep slopes as talus cones and is called colluvial material.

Cumulose parent material is not moved but originates from deposited organic material. This includes peat and muck soils and results from preservation of plant residues by the low oxygen content of a high water table. While peat may form sterile soils, muck soils may be very fertile.

Weathering of Parent Material

The weathering of parent material takes the form of physical weathering (disintegration), chemical weathering (decomposition) and chemical transformation. Generally, minerals that are formed under the high temperatures and pressures at great depths within the Earth's mantle are less resistant to weathering, while minerals formed at low temperature and pressure environment of the surface are more resistant to weathering. Weathering is usually confined to the top few meters of geologic material, because physical, chemical, and biological stresses generally decrease with depth. Physical disintegration begins as rocks that have solidified deep in the Earth are exposed to lower pressure near the surface and swell and become mechanically unstable. Chemical decomposition is a function of mineral solubility, the rate of which doubles with each 10 °C rise in temperature, but is strongly dependent on water to effect chemical changes. Rocks that will decompose in a few years in tropical climates will remain unaltered for millennia in deserts. Structural changes are the result of hydration, oxidation, and reduction.

- Physical disintegration is the first stage in the transformation of parent material into soil. Temperature fluctuations cause expansion and contraction of the rock, splitting it along lines of weakness. Water may then enter the cracks and freeze and cause the physical splitting of material along a path toward the center of the rock, while temperature gradients within the rock can cause exfoliation of "shells". Cycles of wetting and drying cause soil particles to be abraded to a finer size, as does the physical rubbing of material as it is moved by wind, water, and gravity. Water can deposit within rocks minerals that expand upon drying, thereby stressing the rock. Finally, organisms reduce parent material in size through the action of plant roots or digging on the part of animals.

- Chemical decomposition and structural changes result when minerals are made soluble by water or are changed in structure. The first three of the following list are solubility changes and the last three are structural changes.

1. The solution of salts in water results from the action of bipolar water on ionic salt compounds producing a solution of ions and water, removing those minerals and reducing the rock's integrity.
2. Hydrolysis is the transformation of minerals into polar molecules by the splitting of intervening water. This results in soluble acid-base pairs. For example, the hydrolysis of orthoclase-feldspar transforms it to acid silicate clay and basic potassium hydroxide, both of which are more soluble.
3. In carbonation, the solution of carbon dioxide in water forms carbonic acid. Carbonic acid will transform calcite into more soluble calcium bicarbonate.
4. Hydration is the inclusion of water in a mineral structure, causing it to swell and leaving it more stressed and easily decomposed.
5. Oxidation of a mineral compound is the inclusion of oxygen in a mineral, causing it to increase its oxidation number and swell due to the relatively large size of oxygen, leaving it stressed and more easily attacked by water (hydrolysis) or carbonic acid (carbonation).
6. Reduction the opposite of oxidation, means the removal of oxygen, hence the oxidation number of some part of the mineral is reduced, which occurs when oxygen is scarce. The reduction of minerals leaves them electrically unstable, more soluble and internally stressed and easily decomposed.

Of the above, hydrolysis and carbonation are the most effective.

Saprolite is a particular example of a residual soil formed from the transformation of granite, metamorphic and other types of bedrock into clay minerals. Often called "weathered granite", saprolite is the result of weathering processes that include: hydrolysis, chelation from organic compounds, hydration (the solution of minerals in water with resulting cation and anion pairs) and physical processes that include freezing and thawing. The mineralogical and chemical composition of the primary bedrock material, its physical features, including grain size and degree of consolidation, and the rate and type of weathering transforms the parent material into a different mineral. The texture, pH and mineral constituents of saprolite are inherited from its parent material.

Climate

The principal climatic variables influencing soil formation are effective precipitation (i.e., precipitation minus evapotranspiration) and temperature, both of which affect the rates of chemical, physical, and biological processes. The temperature and moisture both influence the organic matter content of soil through their effects on the balance between plant growth and microbial decomposition. Climate is the dominant factor in soil formation, and soils show the distinctive characteristics of the climate zones in which they form. For every 10 °C rise in temperature, the rates of biochemical reactions more than double. Mineral precipitation and temperature are the primary climatic influences on soil formation. If warm temperatures and abundant water are present in the profile at the same time, the processes of weathering, leaching, and plant growth will be maximized. Humid climates favor the growth of trees. In contrast, grasses are the dominant native vegetation in subhumid and semiarid regions, while shrubs and brush of various kinds dominate in arid areas.

Water is essential for all the major chemical weathering reactions. To be effective in soil formation, water must penetrate the regolith. The seasonal rainfall distribution, evaporative losses, site topography, and soil permeability interact to determine how effectively precipitation can influence soil formation. The greater the depth of water penetration, the greater the depth of weathering of the soil and its development. Surplus water percolating through the soil profile transports soluble and suspended materials from the upper to the lower layers. It may also carry away soluble materials in the surface drainage waters. Thus, percolating water stimulates weathering reactions and helps differentiate soil horizons. Likewise, a deficiency of water is a major factor in determining the characteristics of soils of dry regions. Soluble salts are not leached from these soils, and in some cases they build up to levels that curtail plant growth. Soil profiles in arid and semi-arid regions are also apt to accumulate carbonates and certain types of expansive clays.

The direct influences of climate include:

- A shallow accumulation of lime in low rainfall areas as caliche
- Formation of acid soils in humid areas
- Erosion of soils on steep hillsides
- Deposition of eroded materials downstream
- Very intense chemical weathering, leaching, and erosion in warm and humid regions where soil does not freeze

Climate directly affects the rate of weathering and leaching. Wind moves sand and smaller particles, especially in arid regions where there is little plant cover. The type and amount of precipitation influence soil formation by affecting the movement of ions and particles through the soil, and aid in the development of different soil profiles. Soil profiles are more distinct in wet and cool climates, where organic materials may accumulate, than in wet and warm climates, where organic materials are rapidly consumed. The effectiveness of water in weathering parent rock material depends on seasonal and daily temperature fluctuations. Cycles of freezing and thawing constitute an effective mechanism which breaks up rocks and other consolidated materials.

Climate also indirectly influences soil formation through the effects of vegetation cover and biological activity, which modify the rates of chemical reactions in the soil.

Topography

The topography, or relief, is characterized by the inclination (slope), elevation, and orientation of the terrain. Topography determines the rate of precipitation or runoff and rate of formation or erosion of the surface soil profile. The topographical setting may either hasten or retard the work of climatic forces.

Steep slopes encourage rapid soil loss by erosion and allow less rainfall to enter the soil before running off and hence, little mineral deposition in lower profiles. In semi-arid regions, the lower effective rainfall on steeper slopes also results in less complete vegetative cover, so there is less plant contribution to soil formation. For all of these reasons, steep slopes prevent the formation of soil from getting very far ahead of soil destruction. Therefore, soils on steep terrain tend to have rather shallow, poorly developed profiles in comparison to soils on nearby, more level sites.

In swales and depressions where runoff water tends to concentrate, the regolith is usually more deeply weathered and soil profile development is more advanced. However, in the lowest landscape positions, water may saturate the regolith to such a degree that drainage and aeration are restricted. Here, the weathering of some minerals and the decomposition of organic matter are retarded, while the loss of iron and manganese is accelerated. In such low-lying topography, special profile features characteristic of wetland soils may develop. Depressions allow the accumulation of water, minerals and organic matter and in the extreme, the resulting soils will be saline marshes or peat bogs. Intermediate topography affords the best conditions for the formation of an agriculturally productive soil.

Organisms

Soil is the most abundant ecosystem on Earth, but the vast majority of organisms in soil are microbes, a great many of which have not been described. There may be a population limit of around one billion cells per gram of soil, but estimates of the number of species vary widely from 50,000 per gram to over a million per gram of soil. The total number of organisms and species can vary widely according to soil type, location, and depth.

Plants, animals, fungi, bacteria and humans affect soil formation. Animals, soil mesofauna and micro-organisms mix soils as they form burrows and pores, allowing moisture and gases to move about. In the same way, plant roots open channels in soils. Plants with deep taproots can penetrate many metres through the different soil layers to bring up nutrients from deeper in the profile. Plants with fibrous roots that spread out near the soil surface have roots that are easily decomposed, adding organic matter. Micro-organisms, including fungi and bacteria, effect chemical exchanges between roots and soil and act as a reserve of nutrients.

Humans impact soil formation by removing vegetation cover with erosion as the result. Their tillage also mixes the different soil layers, restarting the soil formation process as less weathered material is mixed with the more developed upper layers.

Earthworms, ants and termites mix the soil as they burrow, significantly affecting soil formation. Earthworms ingest soil particles and organic residues, enhancing the availability of plant nutrients in the material that passes through their bodies. They aerate and stir the soil and increase the stability of soil aggregates, thereby assuring ready infiltration of water. In addition, as ants and termites build mounds, they transport soil materials from one horizon to another.

In general, the mixing of the soil by the activities of animals, sometimes called pedoturbation, tends to undo or counteract the tendency of other soil-forming processes that create distinct horizons. Termites and ants may also retard soil profile development by denuding large areas of soil around their nests, leading to increased loss of soil by erosion. Large animals such as gophers, moles, and prairie dogs bore into the lower soil horizons, bringing materials to the surface. Their tunnels are often open to the surface, encouraging the movement of water and air into the subsurface layers. In localized areas, they enhance mixing of the lower and upper horizons by creating, and later refilling, underground tunnels. Old animal burrows in the lower horizons often become filled with soil material from the overlying A horizon, creating profile features known as crotovinas.

Vegetation impacts soils in numerous ways. It can prevent erosion caused by excessive rain that might result from surface runoff. Plants shade soils, keeping them cooler and slow evaporation of soil moisture, or conversely, by way of transpiration, plants can cause soils to lose moisture. Plants can form new chemicals that can break down minerals and improve the soil structure. The type and amount of vegetation depends on climate, topography, soil characteristics, and biological factors. Soil factors such as density, depth, chemistry, pH, temperature and moisture greatly affect the type of plants that can grow in a given location. Dead plants and fallen leaves and stems begin their decomposition on the surface. There, organisms feed on them and mix the organic material with the upper soil layers; these added organic compounds become part of the soil formation process.

Human activities widely influence soil formation. For example, it is believed that Native Americans regularly set fires to maintain several large areas of prairie grasslands in Indiana and Michigan. In more recent times, human destruction of natural vegetation and subsequent tillage of the soil for crop production has abruptly modified soil formation. Likewise, irrigating an arid region of soil drastically influences the soil-forming factors, as does adding fertilizer and lime to soils of low fertility.

Time

Time is a factor in the interactions of all the above. While a mixture of sand, silt and clay constitute the texture of a soil and the aggregation of those components produces peds, the development of a distinct B horizon marks the development of a soil. With time, soils will evolve features that depend on the interplay of the prior listed soil-forming factors. It takes decades to several thousand years for a soil to develop a profile. That time period depends strongly on climate, parent material, relief, and biotic activity. For example, recently deposited material from a flood exhibits no soil development as there has not been enough time for the material to form a structure that further defines soil. The original soil surface is buried, and the formation process must begin anew for this deposit. Over time the soil will develop a profile that depends on the intensities of biota and climate. While a soil can achieve relative stability of its properties for extended periods, the soil life cycle ultimately ends in soil conditions that leave it vulnerable to erosion. Despite the inevitability of soil retrogression and degradation, most soil cycles are long.

Soil-forming factors continue to affect soils during their existence, even on "stable" landscapes that are long-enduring, some for millions of years. Materials are deposited on top or are blown or washed from the surface. With additions, removals and alterations, soils are always subject to new conditions. Whether these are slow or rapid changes depends on climate, topography and biological activity.

Physical Properties of Soils

The physical properties of soils, in order of decreasing importance, are texture, structure, density, porosity, consistency, temperature, colour and resistivity. Soil texture is determined by the relative proportion of the three kinds of soil mineral particles, called soil separates: sand, silt, and clay. At the next larger scale, soil structures called peds are created from the soil separates when iron oxides, carbonates, clay, silica and humus, coat particles and cause them to adhere into larger, relatively stable secondary structures. Soil density, particularly bulk density, is a measure of soil compaction. Soil porosity consists of the void part of the soil volume and is occupied by gases or water. Soil consistency is the ability of soil to stick together. Soil temperature and colour are self-defining. Resistivity refers to the resistance to conduction of electric currents and affects the rate of corrosion of metal and concrete structures. These properties may vary through the depth of a soil profile. Most of these properties determine the aeration of the soil and the ability of water to infiltrate and to be held within the soil.

Influence of Soil Texture Separates on Some Properties of Soils			
Property/behavior	Sand	Silt	Clay
Water-holding capacity	Low	Medium to high	High
Aeration	Good	Medium	Poor
Drainage rate	High	Slow to medium	Very slow
Soil organic matter level	Low	Medium to high	High to medium
Decomposition of organic matter	Rapid	Medium	Slow
Warm-up in spring	Rapid	Moderate	Slow
Compactability	Low	Medium	High
Susceptibility to wind erosion	Moderate (High if fine sand)	High	Low
Susceptibility to water erosion	Low (unless fine sand)	High	Low if aggregated, otherwise high
Shrink/Swell Potential	Very Low	Low	Moderate to very high
Sealing of ponds, dams, and landfills	Poor	Poor	Good
Suitability for tillage after rain	Good	Medium	Poor
Pollutant leaching potential	High	Medium	Low (unless cracked)
Ability to store plant nutrients	Poor	Medium to High	High
Resistance to pH change	Low	Medium	High

Texture

The mineral components of soil are sand, silt and clay, and their relative proportions determine a soil's texture. Properties that are influenced by soil texture, include porosity, permeability, infiltration, shrink-swell rate, water-holding capacity, and susceptibility to erosion. In the illustrated USDA textural classification triangle, the only soil in which neither sand, silt nor clay predominates is called "loam". While even pure sand, silt or clay may be considered a soil, from the perspective of food production a loam soil with a small amount of organic material is considered ideal. The mineral constituents of a loam soil might be 40% sand, 40% silt and the balance 20% clay by weight. Soil texture affects soil behaviour, in particular its retention capacity for nutrients and water.

Sand and silt are the products of physical and chemical weathering of the parent rock; clay, on the other hand, is a most often the product of the precipitation of the dissolved parent rock as a secondary mineral. It is the surface area to volume ratio (specific surface area) of soil particles and the unbalanced ionic charges within those that determine their role in the fertility of soil, as measured by its cation exchange capacity. Sand is least active, followed by silt; clay is the most active. Sand's greatest benefit to soil is that it resists compaction and increases a soil's porosity. Silt is mineralogically like sand but with its higher specific surface area it is more chemically active than sand. But it is the clay content of soil, with its very high specific surface area and generally large number of negative charges, that gives a soil its high retention capacity for water and nutrients. Clay soils also resist wind and water erosion better than silty and sandy soils, as the particles bond tightly to each other.

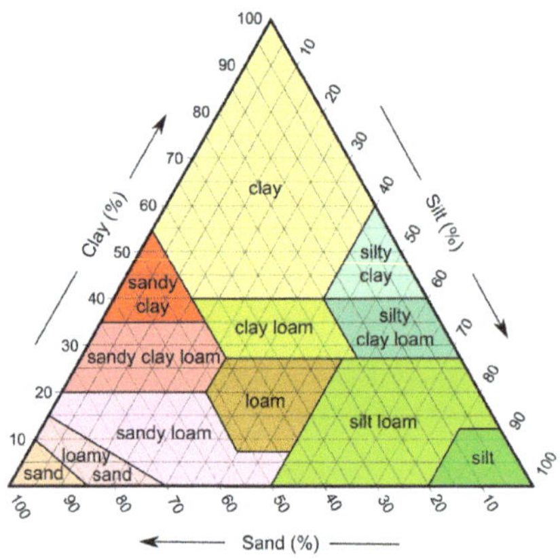

Soil types by clay, silt and sand composition as used by the USDA

Sand is the most stable of the mineral components of soil; it consists of rock fragments, primarily quartz particles, ranging in size from 2.0 to 0.05 mm (0.0787 to 0.0020 in) in diameter. Silt ranges in size from 0.05 to 0.002 mm (0.002 to 0.00008 in). Clay cannot be resolved by optical microscopes as its particles are 0.002 mm (7.9×10^{-5} in) or less in diameter and a thickness of only 10 angstroms (10^{-10} m). In medium-textured soils, clay is often washed downward through the soil profile and accumulates in the subsoil.

Iron-rich soil near Paint Pots in Kootenay National Park, Canada

Soil components larger than 2.0 mm (0.079 in) are classed as rock and gravel and are removed before determining the percentages of the remaining components and the texture class of the soil, but are included in the name. For example, a sandy loam soil with 20% gravel would be called gravelly sandy loam.

When the organic component of a soil is substantial, the soil is called organic soil rather than mineral soil. A soil is called organic if:

1. Mineral fraction is 0% clay and organic matter is 20% or more
2. Mineral fraction is 0% to 50% clay and organic matter is between 20% and 30%
3. Mineral fraction is 50% or more clay and organic matter 30% or more.

Structure

The clumping of the soil textural components of sand, silt and clay causes aggregates to form and the further association of those aggregates into larger units creates soil structures called peds (a contraction of the word pedolith). The adhesion of the soil textural components by organic substances, iron oxides, carbonates, clays, and silica, and the breakage of those aggregates from expansion-contraction, caused by freezing-thawing and wetting-drying cycles, shape soil into distinct geometric forms. The peds evolve into units which have various shapes, sizes and degrees of development. A soil clod, however, is not a ped but rather a mass of soil that results from mechanical disturbance of the soil. Soil structure affects aeration, water movement, conduction of heat, plant root growth and resistance to erosion. Water, in turn, has its strongest effect on soil structure due to its solution and precipitation of minerals and its effect on plant growth.

Soil structure often gives clues to its texture, organic matter content, biological activity, past soil evolution, human use, and the chemical and mineralogical conditions under which the soil formed. While texture is defined by the mineral component of a soil and is an innate property of the soil that does not change with agricultural activities, soil structure can be improved or destroyed by the choice and timing of farming practices.

Soil structural classes:

1. Types: Shape and arrangement of peds
    1. Platy: Peds are flattened one atop the other 1–10 mm thick. Found in the A-horizon of forest soils and lake sedimentation.
    2. Prismatic and Columnar: Prismlike peds are long in the vertical dimension, 10–100 mm wide. Prismatic peds have flat tops, columnar peds have rounded tops. Tend to form in the B-horizon in high sodium soil where clay has accumulated.

    3. Angular and subangular: Blocky peds are imperfect cubes, 5–50 mm, angular have sharp edges, subangular have rounded edges. Tend to form in the B-horizon where clay has accumulated and indicate poor water penetration.
    4. Granular and Crumb: Spheroid peds of polyhedrons, 1–10 mm, often found in the A-horizon in the presence of organic material. Crumb peds are more porous and are considered ideal.

2. Classes: Size of peds whose ranges depend upon the above type
    1. Very fine or very thin: <1 mm platy and spherical; <5 mm blocky; <10 mm prismlike.
    2. Fine or thin: 1–2 mm platy, and spherical; 5–10 mm blocky; 10–20 mm prismlike.
    3. Medium: 2–5 mm platy, granular; 10–20 mm blocky; 20-50 prismlike.
    4. Coarse or thick: 5–10 mm platy, granular; 20–50 mm blocky; 50–100 mm prismlike.
    5. Very coarse or very thick: >10 mm platy, granular; >50 mm blocky; >100 mm prismlike.
3. Grades: Is a measure of the degree of development or cementation within the peds that results in their strength and stability.
    1. Weak: Weak cementation allows peds to fall apart into the three textural constituents, sand, silt and clay.
    2. Moderate: Peds are not distinct in undisturbed soil but when removed they break into aggregates, some broken aggregates and little unaggregated material. This is considered ideal.
    3. Strong:Peds are distinct before removed from the profile and do not break apart easily.
    4. Structureless: Soil is entirely cemented together in one great mass such as slabs of clay or no cementation at all such as with sand.

At the largest scale, the forces that shape a soil's structure result from swelling and shrinkage that initially tend to act horizontally, causing vertically oriented prismatic peds. Clayey soil, due to its differential drying rate with respect to the surface, will induce horizontal cracks, reducing columns to blocky peds. Roots, rodents, worms, and freezing-thawing cycles further break the peds into a spherical shape.

At a smaller scale, plant roots extend into voids and remove water causing the open spaces to increase, thereby decreasing aggregate size. At the same time, roots, fungal hyphae, and earthworms create microscopic tunnels that break up peds.

At an even smaller scale, soil aggregation continues as bacteria and fungi exude sticky polysaccharides which bind soil into smaller peds. The addition of the raw organic matter that bacteria and fungi feed upon encourages the formation of this desirable soil structure.

At the lowest scale, the soil chemistry affects the aggregation or dispersal of soil particles. The clay particles contain polyvalent cations which give the faces of clay layers localized negative charges. At the same time, the edges of the clay plates have a slight positive charge, thereby allowing the edges to adhere to the negative charges on the faces of other clay particles or to flocculate (form clumps). On the other hand, when monovalent ions, such as sodium, invade and displace the polyvalent cations, they weaken the positive charges on the edges, while the negative surface charges are relatively strengthened. This leaves negative charge on the clay faces that repel other clay, causing the particles to push apart, and by doing,the flocculation of clay particles into larger, open assemblages. As a result, the clay disperses and settles into voids between peds, causing those to close. In this way the open structure of the soil is destroyed and the soil is made impenetrable to air and water. Such sodic soil tends to form columnar structures near the surface.

Density

Soil particle density is typically 2.60 to 2.75 grams per cm^3 and is usually unchanging for a given soil. Soil particle density is lower for soils with high organic matter content, and is higher for soils with high iron-oxides content. Soil bulk density is equal to the dry mass of the soil divided by the volume of the soil; i.e., it includes air space and organic materials of the soil volume. The soil bulk density of cultivated loam is about 1.1 to 1.4 g/cm^3 (for comparison water is 1.0 g/cm^3). Soil bulk density is highly variable for a given soil. A lower bulk density by itself does not indicate suitability for plant growth due to the influence of soil texture and structure. A high bulk density is indicative of either soil compaction or high sand content. Soil bulk density is inherently always less than the soil particle density.

Representative bulk densities of soils. The percentage pore space was calculated using 2.7 g/cm^3 for particle density except for the peat soil, which is estimated.

Soil treatment and identification	Bulk density g/cm^3	Pore space %
Tilled surface soil of a cotton field	1.3	51
Trafficked inter-rows where wheels passed surface	1.67	37
Traffic pan at 25 cm deep	1.7	36
Undisturbed soil below traffic pan, clay loam	1.5	43
Rocky silt loam soil under aspen forest	1.62	40

Loamy sand surface soil	1.5	43
Decomposed peat	0.55	65

Porosity

Pore space is that part of the bulk volume of soil that is not occupied by either mineral or organic matter but is open space occupied by either gases or water. In a productive, medium-textured soil the total pore space is typically about 50% of the soil volume. Pore size varies considerably; the smallest pores (cryptopores; <0.1 μm) hold water too tightly for use by plant roots; plant-available water is held in ultramicropores, micropores and mesopores (0.1-75 μm); and macropores (>75 μm) are generally air-filled when the soil is at field capacity.

Soil texture determines total volume of the smallest pores; clay soils have smaller pores, but more total pore space than sands. Soil structure has a strong influence on the larger pores that affect soil aeration, water infiltration and drainage. Tillage has the short-term benefit of temporarily increasing the number of pores of largest size, but these can be rapidly degraded by the destruction of soil aggregation.

The pore size distribution affects the ability of plants and other organisms to access water and oxygen; large, continuous pores allow rapid transmission of air, water and dissolved nutrients through soil, and small pores store water between rainfall or irrigation events. Pore size variation also compartmentalizes the soil pore space such that many microorganisms are not in direct competition with one another, which may explain not only the large number of species present, but the fact that functionally redundant microorganisms (organisms with the same ecological niche) can co-exist within the same soil.

Consistency

Consistency is the ability of soil to stick to itself or to other objects (cohesion and adhesion respectively) and its ability to resist deformation and rupture. It is of approximate use in predicting cultivation problems and the engineering of foundations. Consistency is measured at three moisture conditions: air-dry, moist, and wet. In those conditions the consistency quality depends upon the clay content. In the wet state, the two qualities of stickiness and plasticity are assessed. A soil's resistance to fragmentation and crumbling is assessed in the dry state by rubbing the sample. Its resistance to shearing forces is assessed in the moist state by thumb and finger pressure. Additionally, the cemented consistency depends on cementation by substances other than clay, such as calcium carbonate, silica, oxides and salts; moisture content has little effect on its assessment. The measures of consistency border on subjective compared to other measures such as pH, since they employ the apparent feel of the soil in those states.

The terms used to describe the soil consistency in three moisture states and a last not affected by the amount of moisture are as follows:

1. Consistency of Dry Soil: loose, soft, slightly hard, hard, very hard, extremely hard
2. Consistency of Moist Soil: loose, very friable, friable, firm, very firm, extremely firm
3. Consistency of Wet Soil: nonsticky, slightly sticky, sticky, very sticky; nonplastic, slightly plastic, plastic, very plastic
4. Consistency of Cemented Soil: weakly cemented, strongly cemented, indurated (requires hammer blows to break up)

Soil consistency is useful in estimating the ability of soil to support buildings and roads. More precise measures of soil strength are often made prior to construction.

Temperature

Soil temperature depends on the ratio of the energy absorbed to that lost. Soil has a temperature range between -20 to 60 °C. Soil temperature regulates seed germination, plant and root growth and the availability of nutrients. Below 50 cm (20 in), soil temperature seldom changes and can be approximated by adding 1.8 °C (2 °F) to the mean annual air temperature. Soil temperature has important seasonal, monthly and daily variations. Fluctuations in soil temperature are much lower with increasing soil depth. Heavy mulching (a type of soil cover) can slow the warming of soil, and, at the same time, reduce fluctuations in surface temperature.

Most often, agricultural activities must adapt to soil temperatures by:

1. maximizing germination and growth by timing of planting
2. optimizing use of anhydrous ammonia by applying to soil below 10 °C (50 °F)
3. preventing heaving and thawing due to frosts from damaging shallow-rooted crops
4. preventing damage to desirable soil structure by freezing of saturated soils
5. improving uptake of phosphorus by plants

Soil temperatures can be raised by drying soils or the use of clear plastic mulches. Organic mulches slow the warming of the soil.

There are various factors that affect soil temperature, such as water content, soil color, and relief (slope, orientation, and elevation), and soil cover (shading and insulation). The color of the ground cover and its insulating properties have a strong influence on soil temperature. Whiter soil tends to have a higher albedo than blacker soil cover, which encourages whiter soils to have lower soil temperatures. The specific heat of soil is the energy required to raise the temperature of soil by 1 °C. The specific heat of soil increases as water content increases, since the heat capacity of water is greater than that of dry soil. The specific heat of pure water is ~ 1 calorie per gram, the specific heat of dry soil is ~ 0.2 calories per gram, hence, the specific heat of wet soil is ~ 0.2 to 1 calories per gram. Also, a tremendous energy (~540 cal/g) is required to evaporate water (known as the heat of vaporization). As such, wet soil usually warms more slowly than dry soil - wet surface soil is typically 3 to 6 °C colder than dry surface soil.

Soil heat flux refers to the rate at which heat energy moves through the soil in response to a temperature difference between two points in the soil. The heat flux density is the amount of energy that flows through soil per unit area per unit time and has both magnitude and direction. For the simple case of conduction into or out of the soil in the vertical direction, which is most often applicable the heat flux density is:

$$q_x = -k\frac{\delta T}{\delta x}$$

In SI units

q is the heat flux density, in SI the units are $W{\cdot}m^{-2}$

k is the soils' conductivity, $W{\cdot}m^{-1}{\cdot}K^{-1}$. The thermal conduc-tivity is sometimes a constant, otherwise an average value of conductivity for the soil condition between the surface and the point at depth is used.

δT is the temperature difference (temperature gra-dient) between the two points in the soil between which the heat flux density is to be calculated. In SI the units are kelvin, K.

δx is the distance between the two points within the soil, at which the temperatures are measured and between which the heat flux density is being calculated. In SI the units are meters m, and where x is mea-sured positive downward.

Heat flux is in the direction opposite the temperature gradient, hence the minus sign. That is to say, if the temperature of the surface is higher than at depth x the negative sign will result in a positive value for the heat flux q, and which is interpreted as the heat being conducted into the soil.

Component	Thermal Conductivity (W·m-1·K-1)
Quartz	8.8
Clay	2.9
Organic matter	0.25

Water	0.57
Ice	2.4
Air	0.025
Dry soil	0.2-0.4
Wet soil	1-3

Soil temperature is important for the survival and early growth of seedlings. Soil temperatures affect the anatomical and morphological character of root systems. All physical, chemical, and biological processes in soil and roots are affected in particular because of the increased viscosities of water and protoplasm at low temperatures. In general, climates that do not preclude survival and growth of white spruce above ground are sufficiently benign to provide soil temperatures able to maintain white spruce root systems. In some northwestern parts of the range, white spruce occurs on permafrost sites and although young unlignified roots of conifers may have little resistance to freezing, less than half of the "secondary mature" root system of white spruce was killed by exposure to a temperature of 23.3 °C in multiple year experiment with containerized trees from local nurseries in Massachusetts.

Optimum temperatures for tree root growth range between 10 °C and 25 °C in general and for spruce in particular. In 2-week-old white spruce seedlings that were then grown for 6 weeks in soil at temperatures of 15 °C, 19 °C, 23 °C, 27 °C, and 31 °C; shoot height, shoot dry weight, stem diameter, root penetration, root volume, and root dry weight all reached maxima at 19 °C.

However, whereas strong positive relationships between soil temperature (5 °C to 25 °C) and growth have been found in trembling aspen and balsam poplar, white and other spruce species have shown little or no changes in growth with increasing soil temperature. Such insensitivity to soil low temperature may be common among a number of western and boreal conifers.

Color

Soil colour is often the first impression one has when viewing soil. Striking colours and contrasting patterns are especially noticeable. The Red River of the South carries sediment eroded from extensive reddish soils like Port Silt Loam in Oklahoma. The Yellow River in China carries yellow sediment from eroding loess soils. Mollisols in the Great Plains of North America are darkened and enriched by organic matter. Podsols in boreal forests have highly contrasting layers due to acidity and leaching.

In general, color is determined by the organic matter content, drainage conditions, and degree of oxidation. Soil color, while easily discerned, has little use in predicting soil characteristics. It is of use in distinguishing boundaries within a soil profile, determining the origin of a soil's parent material, as an indication of wetness and waterlogged conditions, and as a qualitative means of measuring organic, salt and carbonate contents of soils. Color is recorded in the Munsell color system as for instance 10YR3/4 *Dusky Red*.

Soil color is primarily influenced by soil mineralogy. Many soil colours are due to various iron minerals. The development and distribution of colour in a soil profile result from chemical and biological weathering, especially redox reactions. As the primary minerals in soil parent material weather, the elements combine into new and colourful compounds. Iron forms secondary minerals of a yellow or red colour, organic matter decomposes into black and brown compounds, and manganese, sulfur and nitrogen can form black mineral deposits. These pigments can produce various colour patterns within a soil. Aerobic conditions produce uniform or gradual colour changes, while reducing environments (anaerobic) result in rapid colour flow with complex, mottled patterns and points of colour concentration.

Resistivity

Soil resistivity is a measure of a soil's ability to retard the conduction of an electric current. The electrical resistivity of soil can affect the rate of galvanic corrosion of metallic structures in contact with the soil. Higher moisture content or increased electrolyte concentration can lower resistivity and increase conductivity, thereby increasing the rate of corrosion. Soil resistivity values typically range from about 2 to 1000 $\Omega \cdot m$, but more extreme values are not unusual.

Soil Water

Water that enters a field is removed from a field by runoff, drainage, evaporation or transpiration. Runoff is the water that flows on the surface to the edge of the field; drainage is the water that flows through the soil downward or toward the edge of the field underground; evaporative water loss from a field is that part of the water that evaporates into the atmosphere directly from the field's surface; transpiration is the loss of water from the field by its evaporation from the plant itself.

Water affects soil formation, structure, stability and erosion but is of primary concern with respect to plant growth. Water is essential to plants for four reasons:

1. It constitutes 80%-95% of the plant's protoplasm.
2. It is essential for photosynthesis.
3. It is the solvent in which nutrients are carried to, into and throughout the plant.
4. It provides the turgidity by which the plant keeps itself in proper position.

In addition, water alters the soil profile by dissolving and re-depositing minerals, often at lower levels, and possibly leaving the soil sterile in the case of extreme rainfall and drainage. In a loam soil, solids constitute half the volume, gas one-quarter of the volume, and water one-quarter of the volume of which only half of which will be available to most plants.

A flooded field will drain the gravitational water under the influence of gravity until water's adhesive and cohesive forces resist further drainage at which point it is said to have reached field capacity. At that point, plants must apply suction to draw water from a soil. When soil becomes too dry, the available water is used up and the remaining moisture is called unavailable water as the plant cannot produce sufficient suction to draw that water in. A plant must produce suction that increases from zero for a flooded field to 1/3 bar at field dry condition (one bar is a little less than one atmosphere pressure). At 15 bar suction, wilting percent, seeds will not germinate, plants begin to wilt and then die. Water moves in soil under the influence of gravity, osmosis and capillarity. When water enters the soil, it displaces air from some of the pores, since air content of a soil is inversely related to its water content.

The rate at which a soil can absorb water depends on the soil and its other conditions. As a plant grows, its roots remove water from the largest pores first. Soon the larger pores hold only air, and the remaining water is found only in the intermediate- and smallest-sized pores. The water in the smallest pores is so strongly held to particle surfaces that plant roots cannot pull it away. Consequently, not all soil water is available to plants. When saturated, the soil may lose nutrients as the water drains. Water moves in a drained field under the influence of pressure where the soil is locally saturated and by capillarity pull to dryer parts of the soil. Most plant water needs are supplied from the suction caused by evaporation from plant leaves and 10% is supplied by "suction" created by osmotic pressure differences between the plant interior and the soil water. Plant roots must seek out water. Insufficient water will damage the yield of a crop. Most of the available water is used in transpiration to pull nutrients into the plant.

Water Retention Forces

Water is retained in a soil when the adhesive force of attraction that water's hydrogen atoms have for the oxygen of soil particles is stronger than the cohesive forces that water's hydrogen feels for other water oxygen atoms. When a field is flooded, the soil pore space is completely filled by water. The field will drain under the force of gravity until it reaches what is called field capacity, at which point the smallest pores are filled with water and the largest with water and gases. The total amount of water held when field capacity is reached is a function of the specific surface area of the soil particles. As a result, high clay and high organic soils have higher field capacities. The total force required to pull or push water out of soil is termed suction and usually expressed in units of bars (10^5 pascal) which is just a little less than one-atmosphere pressure. Alternatively, the terms "tension" or "moisture potential" may be used.

Moisture Classification

The forces with which water is held in soils determine its availability to plants. Forces of adhesion hold water strongly to mineral and humus surfaces and less strongly to itself by cohesive forces. A plant's root may penetrate a very small volume of water that is adhering to soil and be initially able to draw in water that is only lightly held by the cohesive forces. But as the droplet is drawn down, the forces of adhesion of the water for the soil particles make reducing the volume of water increasingly difficult until the plant cannot produce sufficient suction to use the remaining water. The remaining water is considered unavailable. The amount of available water depends upon the soil texture and humus amounts and the type of plant attempting to use the water. Cacti, for example, can produce greater suction than can agricultural crop plants.

The following description applies to a loam soil and agricultural crops. When a field is flooded, it is said to be saturated and all available air space is occupied by water. The suction required to draw water into a plant root is zero. As the field drains under the influence of gravity (drained water is called gravitational water or drain-able water), the suction a plant must produce to use such water increases to 1/3 bar. At that point, the soil is said to have reached field capacity, and plants that use the water must produce increasingly higher suction, finally up to 15 bar. At 15 bar suction, the soil water amount is called wilting percent. At that suction the plant cannot sustain its water needs as water is still being lost from the plant by transpiration; the plant's turgidity is lost, and it wilts. The next level, called air-dry, occurs at 1000 bar suction. Finally the oven dry condition is reached at 10,000 bar suction. All water below wilting percentage is called unavailable water.

Soil Moisture Content

When the soil moisture content is optimal for plant growth, the water in the large and intermediate size pores can move about in the soil and be easily used by plants. The amount of water remaining in a soil drained to field capacity and the amount that is available are functions of the soil type. Sandy soil will retain very little water, while clay will hold the maximum amount. The time required to drain a field from flooded condition for a clay loam that begins at 43% water by weight to a field capacity of 21.5% is six days, whereas a sand loam that is flooded to its maximum of 22% water will take two days to reach field capacity of 11.3% water. The available water for the clay loam might be 11.3% whereas for the sand loam it might be only 7.9% by weight.

Wilting point, field capacity, and available water capacity of various soil textures						
Soil Texture	Wilting Point		Field Capacity		Available water capacity	
	Water per foot of soil depth		Water per foot of soil depth		Water per foot of soil depth	
	%	in.	%	in.	%	in.
Medium sand	1.7	0.3	6.8	1.2	5.1	0.9

Fine sand	2.3	0.4	8.5	1.5	6.2	1.1
Sandy loam	3.4	0.6	11.3	2.0	7.9	1.4
Fine sandy loam	4.5	0.8	14.7	2.6	10.2	1.8
Loam	6.8	1.2	18.1	3.2	11.3	2.0
Silt loam	7.9	1.4	19.8	3.5	11.9	2.1
Clay loam	10.2	1.8	21.5	3.8	11.3	2.0
Clay	14.7	2.6	22.6	4.0	7.9	1.4

The above are average values for the soil textures as the percentages of sand, silt and clay vary.

Water Flow in Soils

Water moves through soil due to the force of gravity, osmosis and capillarity. At zero to one-third bar suction, water is pushed through soil from the point of its application under the force of gravity and the pressure gradient created by the pressure of the water; this is called saturated flow. At higher suction, water movement is pulled by capillarity from wetter toward dryer soil. This is caused by water's adhesion to soil solids, and is called unsaturated flow.

Water infiltration and movement in soil is controlled by six factors:

1. Soil texture
2. Soil structure. Fine-textured soils with granular structure are most favourable to infiltration of water.
3. The amount of organic matter. Coarse matter is best and if on the surface helps prevent the destruction of soil structure and the creation of crusts.
4. Depth of soil to impervious layers such as hardpans or bedrock
5. The amount of water already in the soil
6. Soil temperature. Warm soils take in water faster while frozen soils may not be able to absorb depending on the type of freezing.

Water infiltration rates range from 0.25 cm (0.098 in) per hour for high clay soils to 2.5 cm (0.98 in) per hour for sand and well stabilised and aggregated soil structures. Water flows through the ground unevenly, called "gravity fingers", because of the surface tension between water particles. Tree roots create paths for rainwater flow through soil by breaking though soil including clay layers: one study showed roots increasing infiltration of water by 153% and another study showed an increase by 27 times. Flooding temporarily increases soil permeability in river beds, helping to recharge aquifers.

Saturated Flow

Water applied to a soil is pushed by pressure gradients from the point of its application where it is saturated locally, to less saturated areas. Once soil is completely wetted, any more water will move downward, or percolate, carrying with it clay, humus and nutrients, primarily cations, out of the range of plant roots. In order of decreasing solubility, the leached nutrients are:

- Calcium
- Magnesium, Sulfur, Potassium; depending upon soil composition
- Nitrogen; usually little, unless nitrate fertiliser was applied recently
- Phosphorus; very little as its forms in soil are of low solubility.

In the United States percolation water due to rainfall ranges from zero inches just east of the Rocky Mountains to twenty or more inches in the Appalachian Mountains and the north coast of the Gulf of Mexico.

Unsaturated Flow

At suctions less than one-third bar, water moves in all directions via unsaturated flow at a rate that is dependent on the square of the diameter of the water-filled pores. Water is pulled by capillary action due to the adhesion force of water to the soil solids, producing a suction gradient from wet towards drier soil. Doubling the diameter of the pores increases the flow rate by a factor of four. Large pores drained by gravity and not filled with water do not greatly increase the flow rate for unsaturated flow. Water flow is primarily from coarse-textured soil into fine-textured soil and is slowest in fine-textured soils such as clay.

Water Uptake by Plants

Of equal importance to the storage and movement of water in soil is the means by which plants acquire it and their nutrients. Ninety percent of water is taken up by plants as passive absorption caused by the pulling force of water evaporating (transpiring) from the long column of water that leads from the plant's roots to its leaves. In addition, the high concentration of salts within plant roots creates an osmotic pressure gradient that pushes soil water into the roots. Osmotic absorption becomes more important during times of low water transpiration caused by lower temperatures (for example at night) or high humidity. It is the process that causes guttation.

Root extension is vital for plant survival. A study of a single winter rye plant grown for four months in one cubic foot of loam soil showed that the plant developed 13,800,000 roots, a total of 385 miles in length with 2,550 square feet in surface area; and 14 billion hair roots of 6,600 miles total length and 4,320 square feet total area; for a total surface area of 6,870 square feet (83 ft squared). The total surface area of the loam soil was estimated to be 560,000 square feet. In other words, the roots were in contact with only 1.2% of the soil.

Roots must seek out water as the unsaturated flow of water in soil can move only at a rate of up to 2.5 cm (one inch) per day; as a result they are constantly dying and growing as they seek out high concentrations of soil moisture. Insufficient soil moisture, to the point of causing wilting, will cause permanent damage and crop yields will suffer. When grain sorghum was exposed to soil suction as low as 13.0 bar during the seed head emergence through bloom and seed set stages of growth, its production was reduced by 34%.

Consumptive Use and Water Efficiency

Only a small fraction (0.1% to 1%) of the water used by a plant is held within the plant. The majority is ultimately lost via transpiration, while evaporation from the soil surface is also substantial. Transpiration plus evaporative soil moisture loss is called evapotranspiration. Evapotranspiration plus water held in the plant totals to consumptive use, which is nearly identical to evapotranspiration.

The total water used in an agricultural field includes runoff, drainage and consumptive use. The use of loose mulches will reduce evaporative losses for a period after a field is irrigated, but in the end the total evaporative loss will approach that of an uncovered soil. The benefit from mulch is to keep the moisture available during the seedling stage. Water use efficiency is measured by transpiration ratio, which is the ratio of the total water transpired by a plant to the dry weight of the harvested plant. Transpiration ratios for crops range from 300 to 700. For example, alfalfa may have a transpiration ratio of 500 and as a result 500 kilograms of water will produce one kilogram of dry alfalfa.

Soil Atmosphere

The atmosphere of soil is radically different from the atmosphere above. The consumption of oxygen, by microbes and plant roots and their release of carbon dioxide, decrease oxygen and increase carbon dioxide concentration. Atmospheric CO_2 concentration is 0.04%, but in the soil pore space it may range from 10 to 100 times that level. At extreme levels CO_2 is toxic. In addition, the soil voids are saturated with water vapour. Adequate porosity is necessary, not just to allow the penetration of water, but also to allow gases to diffuse in and out. Movement of gases is by diffusion from high concentrations to lower. Oxygen diffuses in and is consumed and excess levels of carbon dioxide, diffuse out with other gases as well as water. Soil texture and structure strongly affect soil porosity and gas diffusion. It is the total pore space (porosity) of soil not the pore size that determines the rate of diffusion of gases into and out of soil. A Platy soil structure and compacted soils (low porosity) impede gas flow, and a deficiency of oxygen may encourage anaerobic bacteria to reduce (strip oxygen) from nitrate NO_3 to the gases N_2, N_2O, and NO, which are then lost to the atmosphere, thereby depleting the soil of nitrogen. Aerated soil is also a net sink of methane CH_4 but a net producer of greenhouse gases when soils are depleted of oxygen and subject to elevated temperatures.

Composition of Soil Particles

Soil particles can be classified by their chemical composition (mineralogy) as well as their size. The particle size distribution of a soil, its texture, determines many of the properties of that soil, but the mineralogy of those particles can strongly modify those properties. The mineralogy of the finest soil particles, clay, is especially important.

Gravel, Sand and Silt

Gravel, sand and silt are the larger soil particles, and their mineralogy is often inherited from the parent material of the soil, but may include products of weathering (such as concretions of calcium carbonate or iron oxide), or residues of plant and animal life (such as silica phytoliths). Quartz is the most common mineral in the sand or silt fraction as it is resistant to chemical weathering; other common minerals are feldspars, micas and ferromagnesian minerals such as pyroxenes, amphiboles and olivines.

Mineral Colloids; Soil Clays

Due to its high specific surface area and its unbalanced negative charges, clay is the most active mineral component of soil. It is a colloidal and most often a crystalline material. In soils, clay is a soil textural class and is defined in a physical sense as any mineral particle less than 2 μm (8×10^{-5} in) in effective diameter. Many soil minerals, such as gypsum, carbonates, or quartz, are small enough to be classified as clay based on their physical size, but chemically they do not afford the same utility as do clay minerals. Chemically, clay is a range of minerals with certain reactive properties.

Clay was once thought to be very small particles of quartz, feldspar, mica, hornblende or augite, but it is now known to be (with the exception of mica-based clays) a precipitate with a mineralogical composition that is dependent on but different from its parent materials and is classed as a secondary mineral. The type of clay that is formed is a function of the parent material and the composition of the minerals in solution. Clay minerals continue to be formed as long as the soil exists. Mica-based clays result from a modification of the primary mica mineral in such a way that it behaves and is classed as a clay. Most clays are crystalline, but some are amorphous. The clays of a soil are a mixture of the various types of clay, but one type predominates.

There are four groups of clay: layer silicates; crystalline chain silicates; metal oxides and hydroxides and oxy-oxides; and amorphous; and allophanes. Most clays are crystalline and most are made up of three or four planes of oxygen held together by planes of aluminium and silicon by way of ionic bonds that together form a single layer of clay. The spatial arrangement of the oxygen atoms determines clay's structure. Half of the weight of clay is oxygen, but on a volume basis oxygen is ninety percent. The layers of clay are sometimes held together through hydrogen bonds or potassium bridges and as a result will swell less in the presence of water. Other clays, such as montmorillonite, have layers that are loosely attached and will swell greatly when water intervenes between the layers.

There are four groups of clays:

1. Layer Crystalline alumino-silica clays: montmorillonite, illite, vermiculite, chlorite, kaolinite.
2. Crystalline Chain carbonate and sulfate minerals: calcite ($CaCO_3$), dolomite ($CaMg(CO_3)_2$) and gypsum ($CaSO_4 \cdot 2H2O$).
3. Amorphous clays: young mixtures of silica (SiO_2-OH) and alumina ($Al(OH)_3$) which have not had time to form regular crystals.
4. Sesquioxide clays: old, highly leached clays which result in oxides of iron, aluminium and titanium.

Alumino-Silica Clays

Alumino-silica clays are characterised by their regular crystalline structure. Oxygen in ionic bonds with silicon forms a tetrahedral coordination (silicon at the center) which in turn forms sheets of silica. Two sheets of silica are bonded together by a plane of aluminium which forms an octahedral coordination, called alumina, with the oxygens of the silica sheet above and that below it. Hydroxyl ions (OH^-) sometimes substitute for oxygen. During the clay formation process, Al^{3+} may substitute for Si^{4+} in the silica layer, and as much as one fourth of the aluminium Al^{3+} may be substituted by Zn^{2+}, Mg^{2+} or Fe^{2+} in the alumina layer. The substitution of lower-valence cations for higher-valence cations (isomorphous substitution) gives clay a local negative charge on an oxygen atom that attracts and holds water and positively charged soil cations, some of which are of value for plant growth. Isomorphous substitution occurs during the clay's formation and does not change with time.

- Montmorillonite clay is made of four planes of oxygen with two silicon and one central aluminium plane intervening. The alumino-silicate montmorillonite clay is said to have a 2:1 ratio of silicon to aluminium. The seven planes together form a single crystal of montmorillonite. The crystals are weakly held together and water may intervene, causing the clay to swell up to ten times its dry volume. It occurs in soils which have had little leaching, hence it is found in arid regions. As the crystals are not bonded face to face, the entire surface is exposed and available for surface reactions, hence it has a high cation exchange capacity (CEC).
- Illite is a 2:1 clay similar in structure to montmorillonite but has potassium bridges between the faces of the clay crystals and the degree of swelling depends on the degree of weathering of the potassium. The active surface area is reduced due to the potassium bonds. Illite originates from the modification of mica, a

primary mineral. It is often found together with montmorillonite and its primary minerals. It has moderate CEC.

- Vermiculite is a mica-based clay similar to illite, but the crystals of clay are held together more loosely by hydrated magnesium and it will swell, but not as much as does montmorillonite. It has very high CEC.
- Chlorite is similar to vermiculite, but the loose bonding by occasional hydrated magnesium, as in vermiculite, is replaced by a hydrated magnesium sheet, that firmly bonds the planes above and below it. It has two planes of silicon, one of aluminium and one of magnesium; hence it is a 2:2 clay. Chlorite does not swell and it has low CEC.
- Kaolinite is very common, highly weathered clay, and more common than montmorillonite in acid soils. It has one silica and one alumina plane per crystal; hence it is a 1:1 type clay. One plane of silica of montmorillonite is dissolved and is replaced with hydroxyls, which produces strong hydrogen bonds to the oxygen in the next crystal of clay. As a result, kaolinite does not swell in water and has a low specific surface area, and as almost no isomorphous substitution has occurred it has a low CEC. Where rainfall is high, acid soils selectively leach more silica than alumina from the original clays, leaving kaolinite. Even heavier weathering results in sesquioxide clays.

silica-sesquioxide

Crystalline Chain Clays

The carbonate and sulfate minerals are much more soluble and hence are found primarily in desert soils where leaching is less active.

Amorphous Clays

Amorphous clays are young, and commonly found in volcanic ash. They are mixtures of alumina and silica which have not formed the ordered crystal shape of alumino-silica clays which time would provide. The majority of their negative charges originates from hydroxyl ions, which can gain or lose a hydrogen ion (H^+) in response to soil pH, in such way was as to buffer the soil pH. They may have either a negative charge provided by the attached hydroxyl ion (OH^-), which can attract a cation, or lose the hydrogen of the hydroxyl to solution and display a positive charge which can attract anions. As a result, they may display either high CEC in an acid soil solution, or high anion exchange capacity in a basic soil solution.

Sesquioxide Clays

Sesquioxide clays are a product of heavy rainfall that has leached most of the silica from alumino-silica clay, leaving the less soluble oxides iron hematite (Fe_2O_3), iron hydroxide ($Fe(OH)_3$), aluminium hydroxide gibbsite ($Al(OH)_3$), hydrated manganese birnessite (MnO_2). It takes hundreds of thousands of years of leaching to create sesquioxide clays. *Sesqui* is Latin for "one and one-half": there are three parts oxygen to two parts iron or aluminium; hence the ratio is one and one-half (not true for all). They are hydrated and act as either amorphous or crystalline. They are not sticky and do not swell, and soils high in them behave much like sand and can rapidly pass water. They are able to hold large quantities of phosphates. Sesquioxides have low CEC but are able to hold anions as well as cations. Such soils range from yellow to red in colour. Such clays tend to hold phosphorus so tightly that it is unavailable for absorption by plants.

Organic Colloids

Humus is the final state of decomposition of organic matter. While it may linger for a thousand years, on the larger scale of the age of the mineral soil components, it is temporary. It is composed of the very stable lignins (30%) and complex sugars (polyuronides, 30%), proteins (30%), waxes, and fats that are resistant to breakdown by microbes. Its chemical assay is 60% carbon, 5% nitrogen, some oxygen and the remainder hydrogen, sulfur, and phosphorus. On a dry weight basis, the CEC of humus is many times greater than that of clay.

Carbon and Terra Preta

In the extreme environment of high temperatures and the leaching caused by the heavy rain of tropical rain forests, the clay and organic colloids are largely destroyed. The heavy rains wash the alumino-silicate clays from the soil leaving only sesquioxide clays of low CEC. The high temperatures and humidity allow bacteria and fungi to virtually dissolve any organic matter on the rain-forest floor overnight and much of the nutrients are volatilized or leached from the soil and lost. However, carbon in the form of charcoal is far more stable than soil colloids and is capable of performing many of the functions of the soil colloids of sub-tropical soils. Soil containing substantial quantities of charcoal, of an anthropogenic origin, is called terra preta. Research into terra preta is still young but is promising. Fallow periods "on the Amazonian Dark Earths can be as short as 6 months, whereas fallow periods on oxisols are usually 8 to 10 years long"

Soil Chemistry

The chemistry of a soil determines its ability to supply available plant nutrients and affects its physical properties and the health of its microbial population. In addition, a soil's chemistry also determines its corrosivity, stability, and ability to absorb pollutants and to filter water. It is the surface chemistry of mineral and organic colloids that determines soil's chemical properties. "A colloid is a small, insoluble, nondiffusible particle larger than a molecule but small enough to remain suspended in a fluid medium without settling. Most soils contain organic colloidal particles called humus as well as the inorganic colloidal particles of clays." The very high specific surface area of colloids and their net charges, gives soil its ability to hold and release ions. Negatively charged sites on colloids attract and release cations in what is referred to as cation exchange. Cation-exchange capacity (CEC) is the amount of exchangeable cations per unit weight of dry soil and is expressed in terms of milliequivalents of positively charged ions per 100 grams of soil (or centimoles of positive charge per kilogram of soil; $cmol_c/kg$). Similarly, positively charged sites on colloids can attract and release anions in the soil giving the soil anion exchange capacity (AEC).

Cation and Anion Exchange

The cation exchange, that takes place between colloids and soil water, buffers (moderates) soil pH, alters soil structure, and purifies percolating water by adsorbing cations of all types, both useful and harmful.

The negative or positive charges on colloid particles make them able to hold cations or anions, respectively, to their surfaces. The charges result from four sources.

1. Isomorphous substitution occurs in clay during its formation, when lower-valence cations substitute for higher-valence cations in the crystal structure. Substitutions in the outermost layers are more effective than for the innermost layers, as the charge strength drops off as the square of the distance. The net result is oxygen atoms with net negative charge and the ability to attract cations.

2. Edge-of-clay oxygen atoms are not in balance ionically as the tetrahedral and octahedral structures are incomplete.

3. Hydroxyls may substitute for oxygens of the silica layers. When the hydrogens of the clay hydroxyls are ionised into solution, they leave the oxygen with a negative charge.

4. Hydrogens of humus hydroxyl groups may be ionised into solution, leaving an oxygen with a negative charge.

Cations held to the negatively charged colloids resist being washed downward by water and out of reach of plants' roots, thereby preserving the fertility of soils in areas of moderate rainfall and low temperatures.

There is a hierarchy in the process of cation exchange on colloids, as they differ in the strength of adsorption by the colloid and hence their ability to replace one another. If present in equal amounts in the soil water solution:

Al^{3+} replaces H^+ replaces Ca^{2+} replaces Mg^{2+} replaces K^+ same as NH^{4+} replaces Na^+

If one cation is added in large amounts, it may replace the others by the sheer force of its numbers. This is called mass action. This is largely what occurs with the addition of fertiliser.

As the soil solution becomes more acidic (low pH, and an abundance of H^+), the other cations more weakly bound to colloids are pushed into solution as hydrogen ions occupy those sites. A low pH may cause hydrogen of hydroxyl groups to be pulled into solution, leaving charged sites on the colloid available to be occupied by other cations. This ionisation of hydroxyl groups on the surface of soil colloids creates what is described as pH-dependent charges. Unlike permanent charges developed by isomorphous substitution, pH-dependent charges are variable and increase with increasing pH. Freed cations can be made available to plants but are also prone to be leached from the soil, possibly making the soil less fertile. Plants are able to excrete H^+ into the soil and by that means, change the pH of the soil near the root and push cations off the colloids, thus making those available to the plant.

Cation Exchange Capacity (CEC)

Cation exchange capacity should be thought of as the soil's ability to remove cations from the soil water solution and sequester those to be exchanged later as the plant roots release hydrogen ions to the solution. CEC is the amount of exchangeable hydrogen cation (H^+) that will combine with 100 grams dry weight of soil and whose measure is one milliequivalents per 100 grams of soil (1 meq/100 g). Hydrogen ions have a single charge and one-thousandth of a gram of hydrogen ions per 100 grams dry soil gives a measure of one milliequivalent of hydrogen ion. Calcium, with an atomic weight 40 times that of hydrogen and with a valence of two, converts to (40/2) x 1 milliequivalent = 20 milliequivalents of hydrogen ion per 100 grams of dry soil or 20 meq/100 g. The modern measure of CEC is expressed as centimoles of positive charge per kilogram (cmol/kg) of oven-dry soil.

Most of the soil's CEC occurs on clay and humus colloids, and the lack of those in hot, humid, wet climates, due to leaching and decomposition respectively, explains the relative sterility of tropical soils. Live plant roots also have some CEC.

Cation exchange capacity for soils; soil textures; soil colloids		
Soil	State	CEC meq/100 g
Charlotte fine sand	Florida	1.0
Ruston fine sandy loam	Texas	1.9
Glouchester loam	New Jersey	11.9
Grundy silt loam	Illinois	26.3
Gleason clay loam	California	31.6
Susquehanna clay loam	Alabama	34.3
Davie mucky fine sand	Florida	100.8
Sands	------	1 - 5
Fine sandy loams	------	5-10
Loams and silt loams	-----	5-15
Clay loams	-----	15-30
Clays	-----	over 30
Sesquioxides	-----	0-3
Kaolinite	-----	3-15
Illite	-----	25-40
Montmorillonite	-----	60-100
Vermiculite (similar to illite)	-----	80-150
Humus	-----	100-300

Anion Exchange Capacity (AEC)

Anion exchange capacity should be thought of as the soil's ability to remove anions from the soil water solution and sequester those for later exchange as the plant roots release carbonate anions to the soil water solution. Those colloids which have low CEC tend to have some AEC. Amorphous and sesquioxide clays have the highest AEC, followed by the iron oxides. Levels of AEC are much lower than for CEC. Phosphates tend to be held at anion exchange sites.

Iron and aluminum hydroxide clays are able to exchange their hydroxide anions (OH^-) for other anions. The order reflecting the strength of anion adhesion is as follows:

$$H_2PO_4^- \text{ replaces } SO_4^{2-} \text{ replaces } NO_3^- \text{ replaces } Cl^-$$

The amount of exchangeable anions is of a magnitude of tenths to a few milliequivalents per 100 g dry soil. As pH rises, there are relatively more hydroxyls, which will displace anions from the colloids and force them into solution and out of storage; hence AEC decreases with increasing pH (alkalinity).

Soil Reaction (pH)

Soil reactivity is expressed in terms of pH and is a measure of the acidity or alkalinity of the soil. More precisely, it is a measure of hydrogen ion concentration in an aqueous solution and ranges in values from 0 to 14 (acidic to basic) but practically speaking for soils, pH ranges from 3.5 to 9.5, as pH values beyond those extremes are toxic to life forms.

Soil pH

At 25 °C an aqueous solution that has a pH of 3.5 has $10^{-3.5}$ moles H^+ (hydrogen ions) per litre of solution (and also $10^{-10.5}$ mole/litre OH^-). A pH of 7, defined as neutral, has 10^{-7} moles hydrogen ions per litre of solution and also 10^{-7} moles of OH^- per litre; since the two concentrations are equal, they are said to neutralise each other. A pH of 9.5 has $10^{-9.5}$ moles hydrogen ions per litre of solution (and also $10^{-2.5}$ mole per litre OH^-). A pH of 3.5 has one million times more hydrogen ions per litre than a solution with pH of 9.5 ($9.5 - 3.5 = 6$ or 10^6) and is more acidic.

The effect of pH on a soil is to remove from the soil or to make available certain ions. Soils with high acidity tend to have toxic amounts of aluminium and manganese. Plants which need calcium need moderate alkalinity, but most minerals are more soluble in acid soils. Soil organisms are hindered by high acidity, and most agricultural crops do best with mineral soils of pH 6.5 and organic soils of pH 5.5.

In high rainfall areas, soils tend to acidity as the basic cations are forced off the soil colloids by the mass action of hydrogen ions from the rain as those attach to the colloids. High rainfall rates can then wash the nutrients out, leaving the soil sterile. Once the colloids are saturated with H^+, the addition of any more hydrogen ions or aluminum hydroxyl cations drives the pH even lower (more acidic) as the soil has been left with no buffering capacity. In areas of extreme rainfall and high temperatures, the clay and humus may be washed out, further reducing the buffering capacity of the soil. In low rainfall areas, unleached calcium pushes pH to 8.5 and with the addition of exchangeable sodium, soils may reach pH 10. Beyond a pH of 9, plant growth is reduced. High pH results in low micro-nutrient mobility, but water-soluble chelates of those nutrients can correct the deficit. Sodium can be reduced by the addition of gypsum (calcium sulphate) as calcium adheres to clay more tightly than does sodium causing sodium to be pushed into the soil water solution where it can be washed out by an abundance of water.

Base Saturation Percentage

There are acid-forming cations (hydrogen and aluminium) and there are base-forming cations. The fraction of the base-forming cations that occupy positions on the soil colloids is called the base saturation percentage. If a soil has a CEC of 20 meq and 5 meq are aluminium and hydrogen cations (acid-forming), the remainder of positions on the colloids (20-5 = 15 meq) are assumed occupied by base-forming cations, so that the percentage base saturation is 15/20 x 100% = 75% (the compliment 25% is assumed acid-forming cations). When the soil pH is 7 (neutral), base saturation is 100 percent and there are no hydrogen ions stored on the colloids. Base saturation is almost in direct proportion to pH (increases with increasing pH). It is of use in calculating the amount of lime needed to neutralise an acid soil. The amount of lime needed to neutralize a soil must take account of the amount of acid forming ions on the colloids not just those in the soil water solution. The addition of enough lime to neutralize the soil water solution will be insufficient to change the pH, as the acid forming cations stored on the soil colloids will tend to restore the original pH condition as they are pushed off those colloids by the calcium of the added lime.

Buffering of Soils

The resistance of soil to changes in pH, as a result of the addition of acid or basic material, is a measure of the buffering capacity of a soil and increases as the CEC increases. Hence, pure sand has almost no buffering ability, while soils high in colloids have high buffering capacity. Buffering occurs by cation exchange and neutralisation.

The addition of a small amount highly basic aqueous ammonia to a soil will cause the ammonium to displace hydrogen ions from the colloids, and the end product is water and colloidally fixed ammonium, but no permanent change overall in soil pH.

The addition of a small amount of lime, $CaCO_3$, will displace hydrogen ions from the soil colloids, causing the fixation of calcium to colloids and the evolution of CO_2 and water, with no permanent change in soil pH.

The addition of carbonic acid (the solution of CO_2 in water) will displace calcium from colloids, as hydrogen ions are fixed to the colloids, evolving water and slightly alkaline (temporary increase in pH) highly soluble calcium bicarbonate, which will then precipitate as lime ($CaCO_3$) and water at a lower level in the soil profile, with the result of no permanent change in soil pH.

All of the above are examples of the buffering of soil pH. The general principal is that an increase in a particular cation in the soil water solution will cause that cation to be fixed to colloids (buffered) and a decrease in solution of that cation will cause it to be withdrawn from the colloid and moved into solution (buffered). The degree of buffering is limited by the CEC of the soil; the greater the CEC, the greater the buffering capacity of the soil.

Nutrients

Sixteen elements or nutrients are essential for plant growth and reproduction. They are carbon C, hydrogen H, oxygen O, nitrogen N, phosphorus P, potassium K, sulfur S, calcium Ca, magnesium Mg, iron Fe, boron B, manganese Mn, copper Cu, zinc Zn, molybdenum Mo, and chlorine Cl. Nutrients required for plants to complete their life cycle are considered essential nutrients. Nutrients that enhance the growth of plants but are not necessary to complete the plant's life cycle are considered non-essential. With the exception of carbon, hydrogen and oxygen, which are supplied by carbon dioxide and water, the nutrients derive originally from the mineral component of the soil.

Plant uptake of nutrients can only proceed when they are present in a plant-available form. In most situations, nutrients are absorbed in an ionic form from (or together with) soil water. Although minerals are the origin of most nutrients, and the bulk of most nutrient elements in the soil is held in crystalline form within primary and secondary minerals, they weather too slowly to support rapid plant growth. For example, The application of finely ground minerals, feldspar and apatite, to soil seldom provides the necessary amounts of potassium and phosphorus at a rate sufficient for good plant growth, as most of the nutrients remain bound in the crystals of those minerals.

The nutrients adsorbed onto the surfaces of clay colloids and soil organic matter provide a more accessible reservoir of many plant nutrients (e.g. K, Ca, Mg, P, Zn). As plants absorb the nutrients from the soil water, the soluble pool is replenished from the surface-bound pool. The decomposition of soil organic matter by microorganisms is another mechanism whereby the soluble pool of nutrients is replenished - this is important for the supply of plant-available N, S, P, and B from soil.

Gram for gram, the capacity of humus to hold nutrients and water is far greater than that of clay minerals. All in all, small amounts of humus may remarkably increase the soil's capacity to promote plant growth.

Plant nutrients, their chemical symbols, and the ionic forms common in soils and available for plant uptake		
Element	Symbol	Ion or molecule
Carbon	C	CO_2 (mostly through leaves)
Hydrogen	H	H^+, HOH (water)
Oxygen	O	O^{2-}, OH^-, CO_3^{2-}, SO_4^{2-}, CO_2
Phosphorus	P	$H_2PO_4^-$, HPO_4^{2-} (phosphates)
Potassium	K	K^+
Nitrogen	N	NH_4^+, NO_3^- (ammonium, nitrate)
Sulfur	S	SO_4^{2-}
Calcium	Ca	Ca^{2+}
Iron	Fe	Fe^{2+}, Fe^{3+} (ferrous, ferric)
Magnesium	Mg	Mg^{2+}
Boron	B	H_3BO_3, $H_2BO_3^-$, $B(OH)_4^-$
Manganese	Mn	Mn^{2+}
Copper	Cu	Cu^{2+}
Zinc	Zn	Zn^{2+}

Plant nutrients, their chemical symbols, and the ionic forms common in soils and available for plant uptake		
Element	Symbol	Ion or molecule
Molybdenum	Mo	MoO_4^{2-} (molybdate)
Chlorine	Cl	Cl^- (chloride)

Soil Processes Important for Nutrient Uptake

Nutrients in the soil are taken up by the plant through its roots. To be taken up by a plant, a nutrient element must be located near the root surface; however, the supply of nutrients in contact with the root is rapidly depleted. There are three basic mechanisms whereby nutrient ions dissolved in the soil solution are brought into contact with plant roots:

1. Mass flow of water
2. Diffusion within water
3. Interception by root growth

All three mechanisms operate simultaneously, but one mechanism or another may be most important for a particular nutrient. For example, in the case of calcium, which is generally plentiful in the soil solution, mass flow alone can usually bring sufficient amounts to the root surface. However, in the case of phosphorus, diffusion is needed to supplement mass flow. For the most part, nutrient ions must travel some distance in the soil solution to reach the root surface. This movement can take place by mass flow, as when dissolved nutrients are carried along with the soil water flowing toward a root that is actively drawing water from the soil. In this type of movement, the nutrient ions are somewhat analogous to leaves floating down a stream. In addition, nutrient ions continually move by diffusion from areas of greater concentration toward the nutrient-depleted areas of lower concentration around the root surface. That process is due to random motion of molecules. By this means, plants can continue to take up nutrients even at night, when water is only slowly absorbed into the roots as transpiration has almost stopped. Finally, root interception comes into play as roots continually grow into new, undepleted soil.

Estimated relative importance of mass flow, diffusion and root interception as mechanisms in supplying plant nutrients to corn plant roots in soils			
Nutrient	Approximate percentage supplied by:		
	Mass flow	Root interception	Diffusion
Nitrogen	98.8	1.2	0
Phosphorus	6.3	2.8	90.9
Potassium	20.0	2.3	77.7
Calcium	71.4	28.6	0

Sulfur	95.0	5.0	0
Molybdenum	95.2	4.8	0

In the above table, phosphorus and potassium nutrients move more by diffusion than they do by mass flow in the soil water solution, as they are rapidly taken up by the roots creating a concentration of almost zero near the roots (the plants cannot transpire enough water to draw more of those nutrients near the roots). The very steep concentration gradient is of greater influence in the movement of those ions than is the movement of those by mass flow. The movement by mass flow requires the transpiration of water from the plant causing water and solution ions to also move toward the roots. Movement by root interception is slowest as the plants must extend their roots.

Plants move ions out of their roots in an effort to move nutrients in from the soil. Hydrogen H^+ is exchanged for other cations, and carbonate (HCO_3^-) and hydroxide (OH^-) anions are exchanged for nutrient anions. As plant roots remove nutrients from the soil water solution, they are replenished as other ions move off of clay and humus (by ion exchange or desorption), are added from the weathering of soil minerals, and are released by the decomposition of soil organic matter. Plants derive a large proportion of their anion nutrients from decomposing organic matter, which typically holds about 95 percent of the soil nitrogen, 5 to 60 percent of the soil phosphorus and about 80 percent of the soil sulfur. Where crops are produced, the replenishment of nutrients in the soil must usually be augmented by the addition of fertilizer or organic matter.

Because nutrient uptake is an active metabolic process, conditions that inhibit root metabolism may also inhibit nutrient uptake. Examples of such conditions include waterlogging or soil compaction resulting in poor soil aeration, excessively high or low soil temperatures, and above-ground conditions that result in low translocation of sugars to plant roots.

Carbon

Measuring soil respiration in the field using an SRS2000 system.

Plants obtain their carbon from atmospheric carbon dioxide. About 45% of a plant's dry mass is carbon; plant residues typically have a carbon to nitrogen ratio (C/N) of between 13:1 and 100:1. As the soil organic material is digested by arthropods and micro-organisms, the C/N decreases as the carbonaceous material is metabolized and carbon dioxide (CO_2) is released as a byproduct which then finds its way out of the soil and into the atmosphere. The nitrogen is sequestered in the bodies of the living matter of those decomposing organisms and so it builds up in the soil. Normal CO_2 concentration in the atmosphere is 0.03%, this can be the factor limiting plant growth. In a field of maize on a still day during high light conditions in the growing season, the CO_2 concentration drops very low, but under such conditions the crop could use up to 20 times the normal concentration. The respiration of CO_2 by soil micro-organisms decomposing soil organic matter contributes an important amount of CO_2 to the photosynthesising plants. Within the soil, CO_2 concentration is 10 to 100 times that of atmospheric levels but may rise to toxic levels if the soil porosity is low or if diffusion is impeded by flooding.

Nitrogen

Nitrogen is the most critical element obtained by plants from the soil and nitrogen deficiency often limits plant growth. Plants can use the nitrogen as either the ammonium cation (NH_4^+) or the anion nitrate (NO_3^-). Usually, most of the nitrogen in soil is bound within organic compounds that make up the soil organic matter, and must be mineralized to the ammonium or nitrate form before it can be taken up by most plants. The total nitrogen content depends largely on the soil organic matter content, which in turn depends on the climate, vegetation, topography, age and soil management. Soil nitrogen typically decreases by 0.2 to 0.3% for every temperature increase by 10 °C. Usually, grassland soils contain more soil nitrogen than forest soils. Cultivation decreases soil nitrogen by exposing soil organic matter to decomposition by microorganisms, and soils under no-tillage maintain more soil nitrogen than tilled soils.

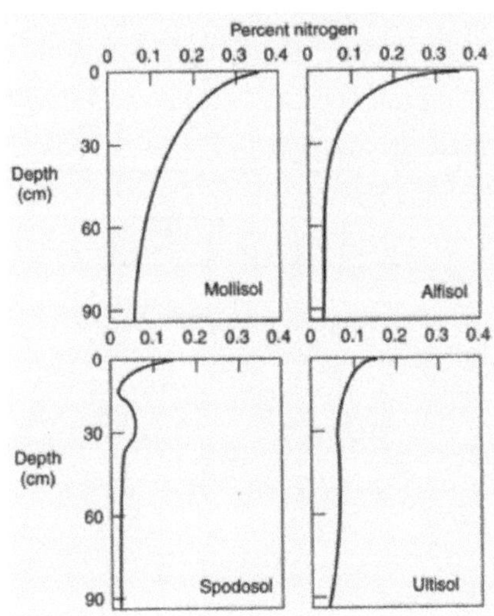

Generalization of percent soil nitrogen by soil order

Some micro-organisms are able to metabolise organic matter and release ammonium in a process called *mineralisation*. Others take free ammonium and oxidise it to nitrate. Nitrogen-fixing bacteria are capable of metabolising N_2 into the form of ammonia in a process called nitrogen fixation. Both ammonium and nitrate can be *immobilized* by their incorporation into the microbes' living cells, where it is temporarily sequestered in the form of amino acids and protein. Nitrate may also be lost from the soil when bacteria metabolise it to the gases N_2 and N_2O. The loss of gaseous forms of nitrogen to the atmosphere due to microbial action is called *denitrification*. Nitrogen may also be *leached* from the soil if it is in the form of nitrate or lost to the atmosphere as ammonia due to a chemical reaction of ammonium with alkaline soil by way of a process called *volatilisation*. Ammonium may also be sequestered in clay by *fixation*. A small amount of nitrogen is added to soil by rainfall.

Nitrogen Gains

In the process of mineralisation, microbes feed on organic matter, releasing ammonia (NH_3), ammonium (NH_4^+) and other nutrients. As long as the carbon to nitrogen ratio (C/N) of fresh residues in the soil is above 30:1, nitrogen will be in short supply and other bacteria will feed on the ammonium and incorporate its nitrogen into their cells in the immobilization process. In that form the nitrogen is said to be *immobilised*. Later, when such bacteria die, they too are *mineralised* and some of the nitrogen is released as ammonium and nitrate. If the C/N is less than 15, ammonia is freed to the soil, where it may be used by bacteria which oxidise it to nitrate (nitrification). Bacteria may on average add 25 pounds (11 kg) nitrogen per acre, and in an unfertilised field, this is the most important source of usable nitrogen. In a soil with 5% organic matter perhaps 2 to 5% of that is released to the soil by such decomposition. It occurs fastest in warm, moist, well aerated soil. The mineralisation of 3% of the organic material of a soil that is 4% organic matter overall, would release 120 pounds (54 kg) of nitrogen as ammonium per acre.

Carbon/Nitrogen Ratio of Various Organic Materials	
Organic Material	C:N Ratio
Alfalfa	13
Bacteria	4
Clover, green sweet	16
Clover, mature sweet	23
Fungi	9
Forest litter	30
Humus in warm cultivated soils	11
Legume-grass hay	25
Legumes (alfalfa or clover), mature	20
Manure, cow	18

Carbon/Nitrogen Ratio of Various Organic Materials	
Organic Material	C:N Ratio
Manure, horse	16–45
Manure, human	10
Oat straw	80
Straw, cornstalks	90
Sawdust	250

In nitrogen fixation, rhizobium bacteria convert N_2 to ammonia (NH_3). Rhizobia share a symbiotic relationship with host plants, since rhizobia supply the host with nitrogen and the host provides rhizobia with nutrients and a safe environment. It is estimated that such symbiotic bacteria in the root nodules of legumes add 45 to 250 pounds of nitrogen per acre per year, which may be sufficient for the crop. Other, free-living nitrogen-fixing bacteria and blue-green algae live independently in the soil and release nitrate when their dead bodies are converted by way of mineralisation.

Some amount of usable nitrogen is fixed by lightning as nitric oxide (NO) and nitrogen dioxide (NO_2^-). Nitrogen dioxide is soluble in water to form nitric acid (HNO_3) solution of H^+ and NO_3^-. Ammonia, NH_3, previously released from the soil or from combustion, may fall with precipitation as nitric acid at a rate of about five pounds nitrogen per acre per year.

Nitrogen Sequestration

When bacteria feed on soluble forms of nitrogen (ammonium and nitrate), they temporarily sequester that nitrogen in their bodies in a process called *immobilisation*. At a later time when those bacteria die, their nitrogen may be released as ammonium by the processes of mineralisation.

Protein material is easily broken down, but the rate of its decomposition is slowed by its attachment to the crystalline structure of clay and when trapped between the clay layers. The layers are small enough that bacteria cannot enter. Some organisms can exude extracellular enzymes that can act on the sequestered proteins. However, those enzymes too may be trapped on the clay crystals.

Ammonium fixation occurs when ammonium pushes potassium ions from between the layers of clay such as illite or montmorillonite. Only a small fraction of soil nitrogen is held this way.

Nitrogen Losses

Usable nitrogen may be lost from soils when it is in the form of nitrate, as it is easily leached. Further losses of nitrogen occur by denitrification, the process whereby soil bacteria convert nitrate (NO_3^-) to nitrogen gas, N_2 or N_2O. This occurs when poor soil aeration limits free oxygen, forcing bacteria to use the oxygen in nitrate for their respiratory process. Denitrification increases when oxidisable organic material is available and when soils are warm and slightly acidic. Denitrification may vary throughout a soil as the aeration varies from place to place. Denitrification may cause the loss of 10 to 20 percent of the available nitrates within a day and when conditions are favourable to that process, losses of up to 60 percent of nitrate applied as fertiliser may occur.

Ammonium volatilisation occurs when ammonium reacts chemically with an alkaline soil, converting NH_4^+ to NH_3. The application of ammonium fertiliser to such a field can result in volatilisation losses of as much as 30 percent.

Phosphorus

After nitrogen, phosphorus is probably the element most likely to be deficient in soils. The soil mineral apatite is the most common mineral source of phosphorus. While there is on average 1000 lb of phosphorus per acre in the soil, it is generally unavailable in the form of phosphates of low solubility. Total phosphorus is about 0.1 percent by weight of the soil, but only one percent of that is available. Of the part available, more than half comes from the mineralisation of organic matter. Agricultural fields may need to be fertilised to make up for the phosphorus that has been removed in the crop.

When phosphorus does form solubilised ions of $H_2PO_4^-$, they rapidly form insoluble phosphates of calcium or hydrous oxides of iron and aluminum. Phosphorus is largely immobile in the soil and is not leached but actually builds up in the surface layer if not cropped. The application of soluble fertilisers to soils may result in zinc deficiencies as zinc phosphates form. Conversely, the application of zinc to soils may immobilise phosphorus again as zinc phosphate. Lack of phosphorus may interfere with the normal opening of the plant leaf stomata, resulting in plant temperatures 10 percent higher than normal. Phosphorus is most available when soil pH is 6.5 in mineral soils and 5.5 in organic soils.

Potassium

The amount of potassium in a soil may be as much as 80,000 lb per acre-foot, of which only 150 lb is available for plant growth. Common mineral sources of potassium are the mica biotite and potassium feldspar, $KAlSi_3O_8$. When solubilised, half will be held as exchangeable cations on clay while the other half is in the soil water solution. Potassium fixation often occurs when soils dry and the potassium is bonded between layers of illite clay. Under certain conditions, dependent on the soil texture, intensity of drying, and initial amount of exchangeable potassium, the fixed percentage may be as much as 90 percent within ten minutes. Potassium may be leached from soils low in clay.

Calcium

Calcium is one percent by weight of soils and is generally available but may be low as it is soluble and can be leached. It is thus low in sandy and heavily leached soil or strongly acidic mineral soil. Calcium is supplied to the plant in the form of exchangeable ions and moderately soluble minerals. Calcium is more available on the soil colloids than is potassium because the common mineral calcite, $CaCO_3$, is more soluble than potassium-bearing minerals.

Magnesium

Magnesium is one of the dominant exchangeable cations in most soils (as are calcium and potassium). Primary minerals that weather to release magnesium include hornblende, biotite and vermiculite. Soil magnesium concentrations are generally sufficient for optimal plant growth, but highly weathered and sandy soils may be magnesium deficient due to leaching by heavy precipitation.

Sulfur

Most sulfur is made available to plants, like phosphorus, by its release from decomposing organic matter. Deficiencies may exist in some soils (especially sandy soils) and if cropped, sulfur needs to be added. The application of large quantities of nitrogen to fields that have marginal amounts of sulfur may cause sulfur deficiency in the rapidly growing plants by the plant's growth outpacing the supply of sulfur. A 15-ton crop of onions uses up to 19 lb of sulfur and 4 tons of alfalfa uses 15 lb per acre. Sulfur abundance varies with depth. In a sample of soils in Ohio, United States, the sulfur abundance varied with depths, 0-6 inches, 6-12 inches, 12-18 inches, 18-24 inches in the amounts: 1056, 830, 686, 528 lb per acre respectively.

Micronutrients

The micronutrients essential for plant life, in their order of importance, include iron, manganese, zinc, copper, boron, chlorine and molybdenum. The term refers to plants' needs, not to their abundance in soil. They are required in very small amounts but are essential to plant health in that most are required parts of some enzyme system which speeds up plants' metabolisms. They are generally available in the mineral component of the soil, but the heavy application of phosphates can cause a deficiency in zinc and iron by the formation of insoluble zinc and iron phosphates. Iron deficiency may also result from excessive amounts of heavy metals or calcium minerals (lime) in the soil. Excess amounts of soluble boron, molybdenum and chloride are toxic.

Non-Essential Nutrients

Nutrients which enhance the health but whose deficiency does not stop the life cycle of plants include: cobalt, strontium, vanadium, silicon and nickel. As their importance are evaluated they may be added to the list of essential plant nutrients.

Soil Organic Matter

Soil organic matter is made up of organic compounds and includes plant, animal and microbial material, both living and dead. A typical soil has a biomass composition of 70% microorganisms, 22% macrofauna, and 8% roots. The living component of an acre of soil may include 900 lb of earthworms, 2400 lb of fungi, 1500 lb of bacteria, 133 lb of protozoa and 890 lb of arthropods and algae.

A small part of the organic matter consists of the living cells such as bacteria, molds, and actinomycetes that work to break down the dead organic matter. Were it not for the action of these micro-organisms, the entire carbon dioxide part of the atmosphere would be sequestered as organic matter in the soil.

Chemically, organic matter is classed as follows:

1. Polysaccharides
    1. cellulose
    2. hemicellulose
    3. starch
    4. pectin
2. Lignins
3. Proteins

Most living things in soils, including plants, insects, bacteria, and fungi, are dependent on organic matter for nutrients and/or energy. Soils have organic compounds in varying degrees of decomposition which rate is dependent on the temperature, soil moisture, and aeration. Bacteria and fungi feed on the raw organic matter, which are fed upon by amoebas, which in turn are fed upon by nematodes and arthropods. Organic matter holds soils open, allowing the infiltration of air and water, and may hold as much as twice its weight in water. Many soils, including desert and rocky-gravel soils, have little or no organic matter. Soils that are all organic matter, such as peat (histosols), are infertile. In its earliest stage of decomposition, the original organic material is often called raw organic matter. The final stage of decomposition is called humus.

In grassland, much of the organic matter added to the soil is from the deep, fibrous, grass root systems. By contrast, tree leaves falling on the forest floor are the principal source of soil organic matter in the forest. Another difference is the frequent occurrence in the grasslands of fires that destroy large amounts of aboveground material but stimulate even greater contributions from roots. Also, the much greater acidity under any forests inhibits the action of certain soil organisms that otherwise would mix much of the surface litter into the mineral soil. As a result, the soils under grasslands generally develop a thicker A horizon with a deeper distribution of organic matter than in comparable soils under forests, which characteristically store most of their organic matter in the forest floor (O horizon) and thin A horizon.

Humus

Humus refers to organic matter that has been decomposed by soil flora and fauna to the point where it is resistant to further breakdown. Humus usually constitutes only five percent of the soil or less by volume, but it is an essential source of nutrients and adds important textural qualities crucial to soil health and plant growth. Humus also hold bits of undecomposed organic matter which feed arthropods and worms which further improve the soil. The end product, humus, is soluble in water and forms a weak acid that can attack silicate minerals. Humus is a colloid with a high cation and anion exchange capacity that on a dry weight basis is many times greater than that of clay colloids. It also acts as a buffer, like clay, against changes in pH and soil moisture.

Humic acids and fulvic acids, which begin as raw organic matter, are important constituents of humus. After the death of plants and animals, microbes begin to feed on the residues, resulting finally in the formation of humus. With decomposition, there is a reduction of water-soluble constituents, cellulose and hemicellulose, and nutrients such as nitrogen, phosphorus, and sulfur. As the residues break down, only stable molecules made of aromatic carbon rings, oxygen and hydrogen remain in the form of humin, lignin and lignin complexes collectively called humus. While the structure of humus has few nutrients, it is able to attract and hold cation and anion nutrients by weak bonds that can be released into the soil solution in response to changes in soil pH.

Lignin is resistant to breakdown and accumulates within the soil. It also reacts with amino acids, which further increases its resistance to decomposition, including enzymatic decomposition by microbes. Fats and waxes from plant matter have some resistance to decomposition and persist in soils for a while. Clay soils often have higher organic contents that persist longer than soils without clay as the organic molecules adhere to and are stabilised by the clay. Proteins normally decompose readily, but when bound to clay particles, they become more resistant to decomposition. Clay particles also absorb the enzymes exuded by microbes which would normally break down proteins. The addition of organic matter to clay soils can render that organic matter and any added nutrients inaccessible to plants and microbes for many years. High soil tannin (polyphenol) content can cause nitrogen to be sequestered in proteins or cause nitrogen immobilisation.

Humus formation is a process dependent on the amount of plant material added each year and the type of base soil. Both are affected by climate and the type of organisms present. Soils with humus can vary in nitrogen content but typically have 3 to 6 percent nitrogen. Raw organic matter, as a reserve of nitrogen and phosphorus, is a vital component affecting soil fertility. Humus also absorbs water, and expands and shrinks between dry and wet states, increasing soil porosity. Humus is less stable than the soil's mineral constituents, as it is reduced by microbial decomposition, and over time its concentration diminshes without the addition of new organic matter. However, humus may persist over centuries if not millennia.

Climate and Organic Matter

The production, accumulation and degradation of organic matter are greatly dependent on climate. Temperature, soil moisture and topography are the major factors affecting the accumulation of organic matter in soils. Organic matter tends to accumulate under wet or cold conditions where decomposer activity is impeded by low temperature or excess moisture which results in anaerobic conditions. Conversely, excessive rain and high temperatures of tropical climates enables rapid decomposition of organic matter and leaching of plant nutrients; forest ecosystems on these soils rely on efficient recycling of nutrients and plant matter to maintain their productivity. Excessive slope may encourage the erosion of the top layer of soil which holds most of the raw organic material that would otherwise eventually become humus.

Plant Residue in Soil

Cellulose and hemicellulose undergo fast decomposition by fungi and bacteria, with a half-life of 12–18 days in a temperate climate. Brown rot fungi can decompose the cellulose and hemicellulose, leaving the lignin and phenolic compounds behind. Starch, which is an energy storage system for plants, undergoes fast decomposition by bacteria and fungi. Lignin consists of polymers composed of 500 to 600 units with a highly branched, amorphous structure. Lignin undergoes very slow decomposition, mainly by white rot fungi and actinomycetes; its half-life under temperate conditions is about six months.

Soil Horizons

A horizontal layer of the soil, whose physical features, composition and age are distinct from those above and beneath, are referred to as a soil horizon. The naming of a horizon is based on the type of material of which it is composed. Those materials reflect the duration of specific processes of soil formation. They are labelled using a shorthand notation of letters and numbers which describe the horizon in terms of its colour, size, texture, structure, consistency, root quantity, pH, voids, boundary characteristics and presence of nodules or concretions. Few soil profiles have all the major horizons. Some may have only one horizon.

The exposure of parent material to favourable conditions produces mineral soils that are marginally suitable for plant growth. That growth often results in the accumulation of organic residues. The accumulated organic layer called the O horizon produces a more active soil due to the effect of the organisms that live within it. Organisms colonise and break down organic materials, making available nutrients upon which other plants and animals can live. After sufficient time, humus moves downward and is deposited in a distinctive organic surface layer called the A horizon.

Classification

Soil is classified into categories in order to understand relationships between different soils and to determine the suitability of a soil for a particular use. One of the first classification systems was developed by the Russian scientist Dokuchaev around 1880. It was modified a number of times by American and European researchers, and developed into the system commonly used until the 1960s. It was based on the idea that soils have a particular morphology based on the materials and factors that form them. In the 1960s, a different classification system began to emerge which focused on soil morphology instead of parental materials and soil-forming factors. Since then it has undergone further modifications. The World Reference Base for Soil Resources (WRB) aims to establish an international reference base for soil classification.

Soil Classification Systems

Australia

There are fourteen soil orders at the top level of the Australian Soil Classification. They are: Anthroposols, Organosols, Podosols, Vertosols, Hydrosols, Kurosols, Sodosols, Chromosols, Calcarosols, Ferrosols, Dermosols, Kandosols, Rudosols and Tenosols.

European Union

The EU's soil taxonomy is based on a new standard soil classification in the World Reference Base for Soil Resources produced by the UN's Food and Agriculture Organization. According to this, the major soils in the European Union are:

- Acrisols
- Albeluvisols
- Andosols
- Anthrosols
- Arenosols

- Calcisols
- Cambisols
- Chernozems
- Fluvisols
- Gleysols
- Gypsisols
- Histosols
- Kastanozems
- Leptosols
- Luvisols
- Phaeozems
- Planosols
- Podzols
- Regosols
- Solonchaks
- Solonetzes
- Umbrisols
- Vertisols

USA

A taxonomy is an arrangement in a systematic manner; the USDA soil taxonomy has six levels of classification. They are, from most general to specific: order, suborder, great group, subgroup, family and series. Soil properties that can be measured quantitatively are used in this classification system — they include: depth, moisture, temperature, texture, structure, cation exchange capacity, base saturation, clay mineralogy, organic matter content and salt content. There are 12 soil orders (the top hierarchical level) in the USDA soil taxonomy. The names of the orders end with the suffix *-sol*. The criteria for the different soil orders include properties that reflect major differences in the genesis of soils. The orders are:

- Alfisol - soils with aluminium and iron. They have horizons of clay accumulation, and form where there is enough moisture and warmth for at least three months of plant growth. They constitute 10.1% of soils worldwide.

- Andisol - volcanic ash soils. They are young and very fertile. They cover 1% of the world's ice-free surface.
- Aridisol - dry soils forming under desert conditions which have fewer than 90 consecutive days of moisture during the growing season and are nonleached. They include nearly 12% of soils on Earth. Soil formation is slow, and accumulated organic matter is scarce. They may have subsurface zones of caliche or duripan. Many aridisols have well-developed Bt horizons showing clay movement from past periods of greater moisture.
- Entisol - recently formed soils that lack well-developed horizons. Commonly found on unconsolidated river and beach sediments of sand and clay or volcanic ash, some have an A horizon on top of bedrock. They are 18% of soils worldwide.
- Gelisol - permafrost soils with permafrost within two metres of the surface or gelic materials and permafrost within one metre. They constitute 9.1% of soils worldwide.
- Histosol - organic soils, formerly called bog soils, are 1.2% of soils worldwide.
- Inceptisol - young soils. They have subsurface horizon formation but show little eluviation and illuviation. They constitute 15% of soils worldwide.
- Mollisol - soft, deep, dark fertile soil formed in grasslands and some hardwood forests with very thick A horizons. They are 7% of soils worldwide.
- Oxisol - are heavily weathered, are rich in iron and aluminum oxides (sesquioxides) or kaolin but low in silica. They have only trace nutrients due to heavy tropical rainfall and high temperatures and low CEC of the remaining clays. They are 7.5% of soils worldwide.
- Spodosol - acid soils with organic colloid layer complexed with iron and aluminium leached from a layer above. They are typical soils of coniferous and deciduous forests in cooler climates. They constitute 4% of soils worldwide.
- Ultisol - acid soils in the humid tropics and subtropics, which are depleted in calcium, magnesium and potassium (important plant nutrients). They are highly weathered, but not as weathered as Oxisols. They make up 8% of the soil worldwide.
- Vertisol - inverted soils. They are clay-rich and tend to swell when wet and shrink upon drying, often forming deep cracks into which surface layers can fall. They are difficult to farm or to construct roads and buildings due to their high expansion rate. They constitute 2.4% of soils worldwide.

The percentages listed above are for land area free of ice. "Soils of Mountains", which constitute the balance (11.6%), have a mixture of those listed above, or are classified as "Rugged Mountains" which have no soil.

Spodosol

Vertisol

Inceptisol

Entisol

The above soil orders in sequence of increasing degree of development are Entisols, Inceptisols, Aridisols, Mollisols, Alfisols, Spodosols, Ultisols, and Oxisols. Histosols and Vertisols may appear in any of the above at any time during their development.

The soil suborders within an order are differentiated on the basis of soil properties and horizons which depend on soil moisture and temperature. Forty-seven suborders are recognized in the United States.

The soil great group category is a subdivision of a suborder in which the kind and sequence of soil horizons distinguish one soil from another. About 185 great groups are recognized in the United States. Horizons marked by clay, iron, humus and hard pans and soil features such as the expansion-contraction of clays (that produce self-mixing provided by clay), temperature, and marked quantities of various salts are used as distinguishing features.

The great group categories are divided into three kinds of soil subgroups: typic, intergrade and extragrade. A typic subgroup represents the basic or 'typical' concept of the great group to which the described subgroup belongs. An intergrade subgroup describes the properties that suggest how it grades towards (is similar to) soils of other soil great groups, suborders or orders. These properties are not developed or expressed well enough to cause the soil to be included within the great group towards which they grade, but suggest similarities. Extragrade features are aberrant properties which prevent that soil from being included in another soil classification. About 1,000 soil subgroups are defined in the United States.

A soil family category is a group of soils within a subgroup and describes the physical and chemical properties which affect the response of soil to agricultural management and engineering applications. The principal characteristics used to differentiate soil families include texture, mineralogy, pH, permeability, structure, consistency, the locale's precipitation pattern, and soil temperature. For some soils the criteria also specify the percentage of silt, sand and coarse fragments such as gravel, cobbles and rocks. About 4,500 soil families are recognised in the United States.

A family may contain several soil series which describe the physical location using the name of a prominent physical feature such as a river or town near where the soil sample was taken. An example would be Merrimac for the Merrimack River in New Hampshire, USA. More than 14,000 soil series are recognised in the United States. This permits very specific descriptions of soils.

A soil phase of series, originally called 'soil type' describes the soil surface texture, slope, stoniness, saltiness, erosion, and other conditions.

Uses

Soil is used in agriculture, where it serves as the anchor and primary nutrient base for plants; however, as demonstrated by hydroponics, it is not essential to plant growth if the soil-contained nutrients can be dissolved in a solution. The types of soil and available moisture determine the species of plants that can be cultivated.

Soil material is also a critical component in the mining, construction and landscape development industries. Soil serves as a foundation for most construction projects. The movement of massive volumes of soil can be involved in surface mining, road building and dam construction. Earth sheltering is the architectural practice of using soil for external thermal mass against building walls. Many building materials are soil based.

Soil resources are critical to the environment, as well as to food and fibre production. Soil provides minerals and water to plants. Soil absorbs rainwater and releases it later, thus preventing floods and drought. Soil cleans water as it percolates through it. Soil is the habitat for many organisms: the major part of known and unknown biodiversity is in the soil, in the form of invertebrates (earthworms, woodlice, millipedes, centipedes, snails, slugs, mites, springtails, enchytraeids, nematodes, protists), bacteria, archaea, fungi and algae; and most organisms living above ground have part of them (plants) or spend part of their life cycle (insects) below-ground. Above-ground and below-ground biodiversities are tightly interconnected, making soil protection of paramount importance for any restoration or conservation plan.

The biological component of soil is an extremely important carbon sink since about 57% of the biotic content is carbon. Even on desert crusts, cyanobacteria, lichens and mosses capture and sequester a significant amount of carbon by photosynthesis. Poor farming and grazing methods have degraded soils and released much of this sequestered carbon to the atmosphere. Restoring the world's soils could offset some of the huge increase in greenhouse gases causing global warming, while improving crop yields and reducing water needs.

Waste management often has a soil component. Septic drain fields treat septic tank effluent using aerobic soil processes. Landfills use soil for daily cover. Land application of waste water relies on soil biology to aerobically treat BOD.

Organic soils, especially peat, serve as a significant fuel resource; but wide areas of peat production, such as sphagnum bogs, are now protected because of patrimonial interest.

Geophagy is the practice of eating soil-like substances. Both animals and human cultures occasionally consume soil for medicinal, recreational, or religious purposes. It has been shown that some monkeys consume soil, together with their preferred food (tree foliage and fruits), in order to alleviate tannin toxicity.

Soils filter and purify water and affect its chemistry. Rain water and pooled water from ponds, lakes and rivers percolate through the soil horizons and the upper rock strata, thus becoming groundwater. Pests (viruses) and pollutants, such as persistent organic pollutants (chlorinated pesticides, polychlorinated biphenyls), oils (hydrocarbons), heavy metals (lead, zinc, cadmium), and excess nutrients (nitrates, sulfates, phosphates) are filtered out by the soil. Soil organisms metabolise them or immobilise them in their biomass and necromass, thereby incorporating them into stable humus. The physical integrity of soil is also a prerequisite for avoiding landslides in rugged landscapes.

Degradation

Land degradation refers to a human-induced or natural process which impairs the capacity of land to function. Soils degradation involves the acidification, contamination, desertification, erosion or salination.

Soil acidification is beneficial in the case of alkaline soils, but it degrades land when it lowers crop productivity and increases soil vulnerability to contamination and erosion. Soils are often initially acid because their parent materials were acid and initially low in the basic cations (calcium, magnesium, potassium and sodium). Acidification occurs when these elements are leached from the soil profile by rainfall or by the harvesting of forest or agricultural crops. Soil acidification is accelerated by the use of acid-forming nitrogenous fertilizers and by the effects of acid precipitation.

Soil contamination at low levels is often within a soil's capacity to treat and assimilate waste material. Soil biota can treat waste by transforming it; soil colloids can adsorb the waste material. Many waste treatment processes rely on this treatment capacity. Exceeding treatment capacity can damage soil biota and limit soil function. Derelict soils occur where industrial contamination or other development activity damages the soil to such a degree that the land cannot be used safely or productively. Remediation of derelict soil uses principles of geology, physics, chemistry and biology to degrade, attenuate, isolate or remove soil contaminants to restore soil functions and values. Techniques include leaching, air sparging, chemical amendments, phytoremediation, bioremediation and natural degradation.

Desertification is an environmental process of ecosystem degradation in arid and semi-arid regions, often caused by human activity. It is a common misconception that droughts cause desertification. Droughts are common in arid and semiarid lands. Well-managed lands can recover from drought when the rains return. Soil management tools include maintaining soil nutrient and organic matter levels, reduced tillage and increased cover. These practices help to control erosion and maintain productivity during periods when moisture is available. Continued land abuse during droughts, however, increases land degradation. Increased population and livestock pressure on marginal lands accelerates desertification.

Erosion of soil is caused by water, wind, ice, and movement in response to gravity. More than one kind of erosion can occur simultaneously. Erosion is distinguished from weathering, since erosion also transports eroded soil away from its place of origin (soil in transit may be described as sediment). Erosion is an intrinsic natural process, but in many places it is greatly increased by human activity, especially poor land use practices. These include agricultural activities which leave the soil bare during times of heavy rain or strong winds, overgrazing, deforestation, and improper construction activity. Improved management can limit erosion. Soil conservation techniques which are employed include changes of land use (such as replacing erosion-prone crops with grass or other soil-binding plants), changes to the timing or type of agricultural operations, terrace building, use of erosion-suppressing cover materials (including cover crops and other plants), limiting disturbance during construction, and avoiding construction during erosion-prone periods.

A serious and long-running water erosion problem occurs in China, on the middle reaches of the Yellow River and the upper reaches of the Yangtze River. From the Yellow River, over 1.6 billion tons of sediment flow each year into the ocean. The sediment originates primarily from water erosion (gully erosion) in the Loess Plateau region of northwest China.

Soil piping is a particular form of soil erosion that occurs below the soil surface. It causes levee and dam failure, as well as sink hole formation. Turbulent flow removes soil starting at the mouth of the seep flow and the subsoil erosion advances up-gradient. The term sand boil is used to describe the appearance of the discharging end of an active soil pipe.

Desertification

Soil salination is the accumulation of free salts to such an extent that it leads to degradation of the agricultural value of soils and vegetation. Consequences include corrosion damage, reduced plant growth, erosion due to loss of plant cover and soil structure, and water quality problems due to sedimentation. Salination occurs due to a combination of natural and human-caused processes. Arid conditions favour salt accumulation. This is especially apparent when soil parent material is saline. Irrigation of arid lands is especially problematic. All irrigation water has some level of salinity. Irrigation, especially when it involves leakage from canals and overirrigation in the field, often raises the underlying water table. Rapid salination occurs when the land surface is within the capillary fringe of saline groundwater. Soil salinity control involves watertable control and flushing with higher levels of applied water in combination with tile drainage or another form of subsurface drainage.

Erosion control

Reclamation

Soils which contain high levels of particular clays, such as smectites, are often very fertile. For example, the smectite-rich clays of Thailand's Central Plains are among the most productive in the world.

Many farmers in tropical areas, however, struggle to retain organic matter in the soils they work. In recent years, for example, productivity has declined in the low-clay soils of northern Thailand. Farmers initially responded by adding organic matter from termite mounds, but this was unsustainable in the long-term. Scientists experimented with adding bentonite, one of the smectite family of clays, to the soil. In field trials, conducted by scientists from the International Water Management Institute in cooperation with Khon Kaen University and local farmers, this had the effect of helping retain water and nutrients. Supplementing the farmer's usual practice with a single application of 200 kg bentonite per rai (6.26 rai = 1 hectare) resulted in an average yield increase of 73%. More work showed that applying bentonite to degraded sandy soils reduced the risk of crop failure during drought years.

In 2008, three years after the initial trials, IWMI scientists conducted a survey among 250 farmers in northeast Thailand, half of whom had applied bentonite to their fields. The average improvement for those using the clay addition was 18% higher than for non-clay users. Using the clay had enabled some farmers to switch to growing vegetables, which need more fertile soil. This helped to increase their income. The researchers estimated that 200 farmers in northeast Thailand and 400 in Cambodia had adopted the use of clays, and that a further 20,000 farmers were introduced to the new technique.

If the soil is too high in clay, adding gypsum, washed river sand and organic matter will balance the composition. Adding organic matter (like ramial chipped wood for instance) to soil which is depleted in nutrients and too high in sand will boost its quality.

References

- Buol, S. W.; Hole, F. D.; McCracken, R. J. (1973). Soil Genesis and Classification (1st ed.). Ames, Iowa: Iowa State University Press. ISBN 0-8138-1460-X.
- McCarthy, David F. (1982). Essentials of Soil Mechanics and Foundations: Basic Geotechnics (2nd ed.). Reston, Virginia: Reston Publishing. ISBN 9780835917810.
- Gilluly, James; Waters, Aaron Clement; Woodford, Alfred Oswald (1975). Principles of Geology (4th ed.). W.H. Freeman. ISBN 978-0716702696.
- Gove Hambidge, "Climate and Man—A Summary", in Erwin Raisz, U.S. Department of Agriculture, Climate And Man, Part One, Yearbook of Agriculture 1941, Repr. Honolulu: University Press of the Pacific, 2004, ISBN 978-1-4102-1538-3, pp. 1–66, p. 27.
- Gilluly, Waters, Woodford (1975). Principles of Geology (4th ed.). USA: W.H. Freeman. p. 216. ISBN 978-0716702696.
- Paul, E.A. (1997). Soil organic matter in temperate agroecosystems : long-term experiments in North America. Boca Raton: CRC Press. p. 80. ISBN 9780849328022.
- Buol, S.W. (1990). Soil genesis and classification. Ames, Iowa: Iowa State University Press. p. 36. doi:10.1081/E-ESS. ISBN 0-8138-2873-2.
- IUSS Working Group WRB (2014). World Reference Base for Soil Resources 2014. International soil classification system for naming soils and creating legends for soil maps (PDF) (3rd ed.). Rome: FAO. ISBN 978-92-5-108370-3. Retrieved 29 August 2014.
- Drainage Manual: A Guide to Integrating Plant, Soil, and Water Relationships for Drainage of Irrigated Lands. Interior Dept., Bureau of Reclamation. 1993. ISBN 0-16-061623-9.
- Nyle C. Brady & Ray R. Weil (2009). Elements of the Nature and Properties of Soils (3rd Edition). Prentice Hall. ISBN 9780135014332.
- Danoff-Burg, James A. "The Terrestrial Influence: Geology and Soils". Earth Institute Center for Environmental Sustainability. Columbia University. Retrieved 27 July 2014.
- College of Tropical Agriculture and Human Resources. "Soil Mineralogy". www.ctahr.hawaii.edu. University of Hawaiʻi. Retrieved 7 August 2014.
- Soil Survey Staff (1999). Soil taxonomy: A Basic System of Soil Classification for Making and Interpreting Soil Surveys. 2nd edition. Natural Resources Conservation Service. U.S. Department of

Agriculture Handbook 436. (PDF). USDA: United States Dept. of Agriculture, Naturel Resources Conservation Service. Retrieved 10 March 2013.

- "Arizona Master Gardener Manual". Cooperative Extension, College of Agriculture, University of Arizona. p. Chapter 2, pp 2–4. Retrieved 27 May 2013.
- Blakeslee, Thomas R. (24 February 2010). "Greening Deserts for Carbon Credits". Renewable Energy World.com. Retrieved 23 October 2012.
- Brehm, Denise (11 December 2008). "CEE researchers explain mystery of gravity fingers". MIT Department of Civil & Environmental Engineering. MIT. Retrieved 31 October 2012.
- Copley, Jon (25 August 2005). "Millions of bacterial species revealed underfoot". New Scientist. Retrieved 19 April 2010.
- "Free articles and software on drainage of waterlogged land and soil salinity control". Retrieved 28 July 2010.

3

Properties of Soil

Some of the properties discussed in this chapter are the texture of soil, soil structure, peds, soil color, soil resistivity and soil water. Soil structure is the arrangement of the parts of soil and the pore space located between them. Soil color tells us what is specific about the soil, for instance, soil high in iron is deep orange brown to yellowish brown while the soil that has high content of organic matter is dark brown or black. This chapter is an overview of the subject matter incorporating all the major aspects of soil properties.

Soil Texture

Soil texture is known as a qualitative classification instruments used both the field and laboratory for agricultural soils to determine classes for based on their physical texture. While classes are distinguished in the field and the class is then used to determine crop suitability and to approximate the soils responses to environmental and management conditions such as drought or calcium (lime) requirements. A qualitative rather than a quantitative tool it is a fast, simple and effective means to assess a soil's physical characteristics. The USDA soil taxonomy and WRB soil classification systems use 12 classes whereas the UK-ADAS system uses just 11.

Field Assessment using a Hand Texture Chart

Hand Identification Chart for Soil Texture Analysis

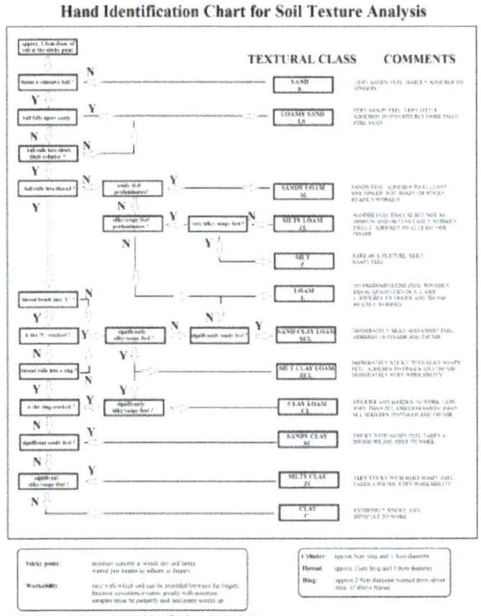

Soil texture chart

Hand analysis, while an arbitrary technique, is an extremely simple and effective means to rapidly assess and classify a soil's physical condition. Correctly executed, the procedure allows for rapid and frequent assessment of soil characteristics with little or no equipment. It is thus an extremely useful tool for identifying spatial variation both within and between plots (fields) as well as identifying progressive changes and boundaries between soil classes and orders.

The method involves taking a small sample of soil, sufficient to roll into a ball of approx 2.5 cm diameter, from just below the surface. Using a small drop of water or 'spit' the sample is then moistened to the sticky point (the point at which it begins to adhere to the finger). The ball is then molded to determine its work ability and its class according to the steps in the chart opposite.

Soil Separates

Soil separates are specific ranges of particle sizes. The smallest particles are *clay* particles and are classified as having diameters of less than 0.002 mm. The next smallest particles are *silt* particles and have diameters between 0.002 mm and 0.05 mm (in USDA soil taxonomy). The largest particles are *sand* particles and are larger than 0.05 mm in diameter. Furthermore, large sand particles can be described as *coarse*, intermediate as *medium*, and the smaller as *fine*. Other countries have their own particle size classifications.

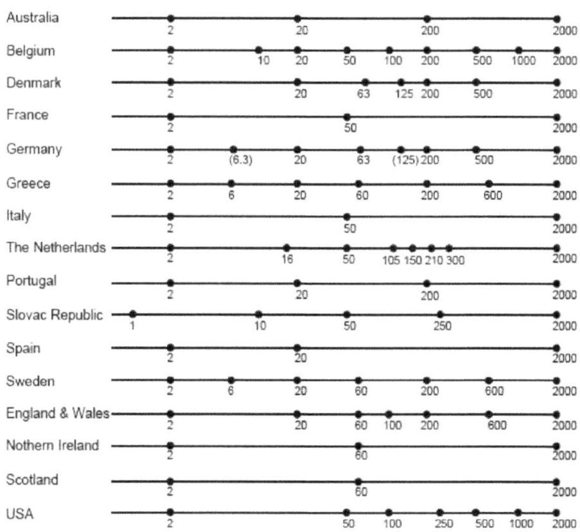

Particle size classifications used by different countries, diameters in μm

Name of soil separate	Diameter limits (mm) (USDA classification)	Diameter limits (mm) (WRB classification)
Clay	less than 0.002	less than 0.002
Silt	0.002–0.05	0.002–0.063

Very fine sand	0.05–0.10	0.063–0.125
Fine sand	0.10–0.25	0.125–0.20
Medium sand	0.25–0.50	0.20–0.63
Coarse sand	0.50–1.00	0.63–1.25
Very coarse sand	1.00-2.00	1.25-2.00

Soil Texture Classification

Soil textures are classified by the fractions of each soil separate (sand, silt, and clay) present in a soil. Classifications are typically named for the primary constituent particle size or a combination of the most abundant particles sizes, e.g. “sandy clay” or “silty clay”. A fourth term, loam, is used to describe a roughly equal concentration of sand, silt, and clay, and lends to the naming of even more classifications, e.g. “clay loam” or “silt loam”.

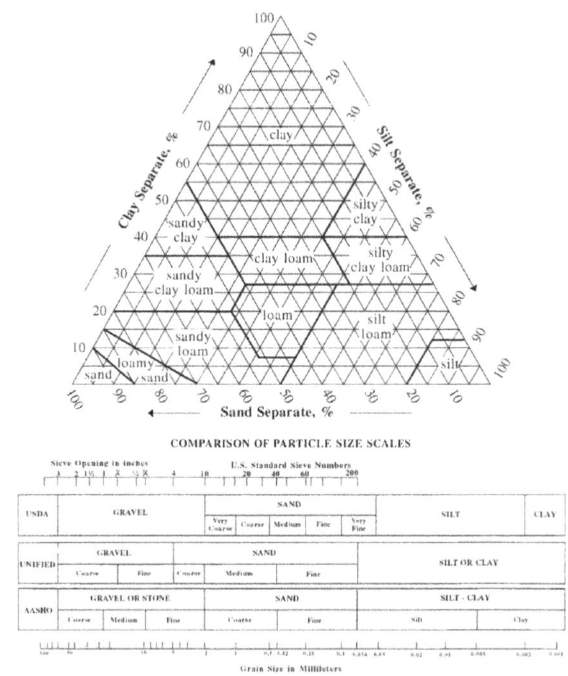

Soil texture triangle, showing the 12 major textural classes, and particle size scales as defined by the USDA.

In the United States, twelve major soil texture classifications are defined by the USDA.

Determining the soil textures is often aided with the use of a soil texture triangle.

History of Classification

The first classification, the International system, was first proposed by Albert Atterberg (1905), and was based on his studies in southern Sweden. Atterberg chose 20 μm for the upper limit of silt fraction because particles smaller than that size were not visible to the naked eye, the suspension could be coagulated by salts, capillary rise within 24 hours was most rapid in this fraction, and the pores between compacted particles were so small as to prevent the entry of root hairs. Commission One of the International Society of Soil Science (ISSS) recommended its use at the first International Congress of Soil Science in Washington in 1927. Australia adopted this system and according to Marshall (1947) its equal logarithmic intervals are an attractive feature worth maintaining. The USDA adopted its own system in 1938, and the FAO used the USDA system in the FAO-UNESCO world soil map and recommended its use.

Soil Structure

Soil structure describes the arrangement of the solid parts of the soil and of the pore space located between them. It is determined by how individual soil granules clump bind together and aggregate, and therefore, the arrangement of soil pores between them. Soil structure has a major influence on water and air movement, biological activity, root growth and seedling emergence.

Overview

Soil structure describes the arrangement of the solid parts of the soil and of the pore spaces located between them (Marshall & Holmes, 1979). The structure depends on what the soil developed from. The quality of soil structure will decline under most forms of cultivation—the associated mechanical mixing of the soil compacts and shears aggregates and fills pore spaces; it also exposes organic matter to a greater rate of decay and oxidation (Young & Young, 2001). A further consequence of continued cultivation and traffic is the development of compacted, impermeable layers or pans within the profile.

The decline of soil structure under irrigation is usually related to the breakdown of aggregates and dispersion of clay material as a result of rapid wetting. This is particularly so if soils are sodic; that is, having a high exchangeable sodium percentage (ESP) of the cations attached to the clays. High sodium levels (compared to high calcium levels) cause particles to repel one another when wet, and the associated aggregates to disaggregate and disperse. The ESP will increase if irrigation causes salty water (even of low concentration) to gain access to the soil.

A wide range of practices are undertaken to preserve and improve soil structure. For example, the NSW Department of Land and Water Conservation, (1991) advocates: increasing organic content by incorporating pasture phases into cropping rotations; reducing or eliminating tillage and cultivation in cropping and pasture activities; avoiding soil disturbance during periods of excessive dry or wet when soils may accordingly tend to shatter or smear; and ensuring sufficient ground cover to protect the soil from raindrop impact. In irrigated agriculture, it may be recommended to: apply gypsum (calcium sulfate) to displace sodium cations with calcium and so reduce ESP or sodicity, avoid rapid wetting, and avoid disturbing soils when too wet or dry.

Impacts of Improving Soil Structure

The benefits of improving soil structure for the growth of plants, particularly in an agricultural setting include: reduced erosion due to greater soil aggregate strength and decreased overland flow; improved root penetration and access to soil moisture and nutrients; improved emergence of seedlings due to reduced crusting of the surface; and greater water infiltration, retention and availability due to improved porosity.

It has been estimated that productivity from irrigated perennial horticulture could be increased by two to three times the present level by improving soil structure, because of the resulting access by plants to available soil water and nutrients (Cockroft & Olsson, 2000, cited in Land and Water Australia 2007). The NSW Department of Land and Water Conservation (1991) suggests that in cropping systems, wheat yields can be increased by 10 kg/ha for every extra millimetre of rain that is able to infiltrate due to soil structure improvements.

Hardsetting Soil

Hardsetting soils lose their structure when wet and then set hard as they dry out to form a structureless mass that is very difficult to cultivate. They can only be tilled when their moisture content is within a limited range. When they are tilled the result is often a very cloddy surface (poor tilth). As they dry out the high soil strength often restricts seedling and root growth. Infiltration rates are low and runoff of rain and irrigation limits the productivity of many hardsetting soils.

Definition

Hardsetting has been defined this way: "*A hardsetting soil is one that sets to an almost homogeneous mass on drying. It may have occasional cracks, typically at a spacing of >0.1 m. Air dry hardset soil is hard and brittle, and it is not possible to push a forefinger into the profile face. Typically, it has a tensile strength of 90 kN* m^{-2}*. Soils that crust are not necessarily hardsetting since a hardsetting horizon is thicker than a crust. (In cultivated soils the thickness of the hardsetting horizon is frequently equal to or greater than that of the cultivated layer.) Hardsetting soil is not permanently cemented and is soft when wet. The clods in a hardsetting horizon that has been cultivated will partially or totally disintegrate upon wetting. If the soil has been sufficiently wetted, it will revert to its hardset state on drying. This can happen after flood irrigation or a single intense rainfall event.*"

Ped

A ped is a soil particle.

There are five major classes of structure seen in soils: platy, prismatic, columnar, granular, and blocky. There are also structureless conditions. Some soils have simple structure, each unit being an entity without component smaller units. Others have compound structure, in which large units are composed of smaller units separated by persistent planes of weakness.

Types of Soil Structures

Platy

In platy structure, the units are flat and platelike. They are generally oriented horizontally. A special form, lenticular platy structure, is recognized for plates that are thickest in the middle and thin toward the edges. Platy structure is usually found in subsurface soils that have been subject to leaching or compaction by animals or machinery. The plates can be separated with a little effort by prying the horizontal layers with a pen knife. Platy structure tends to impede the downward movement of water and plant roots through the soil.

They are found most frequently in the C, E, Bs and K horizons as well as in sesquioxides (very old soils that are rich in iron and magnesium).

Prismatic

In the prismatic structure, the individual units are bounded by flat to rounded vertical faces. Units are distinctly longer vertically, and the faces are typically casts or molds of adjoining units. Vertices are angular or subrounded; the tops of the prisms are somewhat indistinct and normally flat. Prismatic structures are characteristic of the B horizons or subsoils. The vertical cracks result from freezing and thawing and wetting and drying as well as the downward movement of water and roots.

Columnar

In the columnar structure, the units are similar to prisms and are bounded by flat or slightly rounded vertical faces. The tops of columns, in contrast to those of prisms, are very distinct and normally rounded. Columnar structure is common in the subsoil of sodium affected soils and soils rich in swelling clays such as the semectites and the kandite Halloysite. Columnar structure is very dense and it is very difficult for plant roots to penetrate these layers. Techniques such as deep plowing have help to restore some degree of fertility to these soils.

Blocky

In blocky structure, the structural units are blocklike or polyhedral. They are bounded by flat or slightly rounded surfaces that are casts of the faces of surrounding peds. Typically, blocky structural units are nearly equidimensional but grade to prisms and to plates. The structure is described as angular blocky if the faces intersect at relatively sharp angles; as subangular blocky if the faces are a mixture of rounded and plane faces and the corners are mostly rounded. Blocky structures are common in subsoil but also occur in surface soils that have a high clay content. The strongest blocky structure is formed as a result of swelling and shrinking of the clay minerals which produce cracks. Sometimes the surface of dried-up sloughs and ponds shows characteristic cracking and peeling due to clays.

Granular

In the granular structure, the structural units are approximately spherical or polyhedral and are bounded by curved or very irregular faces that are not casts of adjoining peds. In other words, they look like cookie crumbs. Granular structure is common in the surface soils of rich grasslands and highly amended garden soils with high organic matter content. Soil mineral particles are both separated and bridged by organic matter breakdown products, and soil biota exudates, making the soil easy to work. Cultivation, earthworms, frost action and rodents mix the soil and decreases the size of the peds. This structure allows for good porosity and easy movement of air and water. This combination of ease in tillage, good moisture and air handling capabilities, and good structure for planting and germination, are definitive of the phrase *good tilth*.

Pore Space in Soil

The pore space of soil contains the liquid and gas phases of soil, i.e., everything but the solid phase that contains mainly minerals of varying sizes as well as organic compounds.

In order to understand porosity better a series of equations have been used to express the quantitative interactions between the three phases of soil.

Macropores or fractures play a major role in infiltration rates in many soils as well as preferential flow patterns, hydraulic conductivity and evapotranspiration. Cracks are also very influential in gas exchange, influencing respiration within soils. Modeling cracks therefore helps understand how these processes work and what the effects of changes in soil cracking such as compaction, can have on these processes.

Background

Bulk Density

$$\rho = \frac{M_s}{t_V}$$

The bulk density of soil depends greatly on the mineral make up of soil and the degree of compaction. The density of quartz is around 2.65 g/cm^3 but the bulk density of a soil may be less than half that density.

Most soils have a bulk density between 1.0 and 1.6 g/cm^3 but organic soil and some friable clay may have a bulk density well below 1 g/cm^3

Core samples are taken by driving a metal core into the earth at the desired depth and soil horizon. The samples are then oven dried and weighed.

Bulk density = (mass of oven dry soil)/volume

The bulk density of soil is inversely related to the porosity of the same soil. The more pore space in a soil the lower the value for bulk density.

Porosity

$$f = \frac{V_f}{V_t} \text{ or } f = \frac{V_a + V_w}{V_s + V_a + V_w}$$

Porosity is a measure of the total pore space in the soil. This is measured as a volume or percent. The amount of porosity in a soil depends on the minerals that make up the soil and the amount of sorting that occurs within the soil structure. For example, a sandy soil will have larger porosity than silty sand, because the silt will fill in the gaps between the sand particles.

Pore Space Relations

Hydraulic Conductivity

Hydraulic conductivity (K) is a property of soil that describes the ease with which water can move through pore spaces. It depends on the permeability of the material (pores, compaction) and on the degree of saturation. Saturated hydraulic conductivity, K_{sat}, describes water movement through saturated media. Where hydraulic conductivity has the capability to be measured at any state. It can be estimated by numerous kinds of equipment. To calculate hydraulic conductivity, Darcy's law is used. The manipulation of the law depends on the Soil saturation and instrument used.

Infiltration

Infiltration is the process by which water on the ground surface enters the soil. The water enters the soil through the pores by the forces of gravity and capillary action.

The largest cracks and pores offer a great reservoir for the initial flush of water. This allows a rapid infiltration. The smaller pores take longer to fill and rely on capillary forces as well as gravity. The smaller pores have a slower infiltration as the soil becomes more saturated.

Pore Types

A pore is not simply a void in the solid structure of soil. The various pore size categories have different characteristics and contribute different attributes to soils depending on the number and frequency of each type. A widely used classification of pore size is that of Brewer (1964):

Macropore

The pores that are too large to have any significant capillary force. Unless impeded, water will drain from these pores, and they are generally air-filled at field capacity. Macropores can be caused by cracking, division of peds and aggregates, as well as plant roots, and zoological exploration. Size >75 μm.

Mesopore

The largest pores filled with water at field capacity. Also known as storage pores because of the ability to store water useful to plants. They do not have capillary forces too great so that the water does not become limiting to the plants. The properties of mesopores are highly studied by soil scientists because of their impact on agriculture and irrigation. Size 30 μm–75 μm.

Micropore

These are "pores that are sufficiently small that water within these pores is considered immobile, but available for plant extraction." Because there is little movement of water in these pores, solute movement is mainly by the process of diffusion. Size 5-30 μm.

Ultramicropore

These pores are suitable for habitation by microorganisms. Their distribution is determined by soil texture and soil organic matter, and they are not greatly affected by compaction Size 0.1-30 μm.

Cryptopore

Pores that are too small to be penetrated by most microorganisms. Organic matter in these pores is therefore protected from microbial decomposition. They are filled with water unless the soil is very dry, but little of this water is available to plants, and water movement is very slow. Size <0.1 μm.

Modelling Methods

Basic crack modeling has been undertaken for many years by simple observations and measurements of crack size, distribution, continuity and depth. These observations have either been surface observation or done on profiles in pits. Hand tracing and measurement of crack patterns on paper was one method used prior to advances in modern technology. Another field method was with the use of string and a semicircle of wire. The semi circle was moved along alternating sides of a string line. The cracks within the semicircle were measured for width, length and depth using a ruler. The crack distribution was calculated using the principle of Buffon's needle.

Disc Permeameter

This method relies on the fact that crack sizes have a range of different water potentials. At zero water potential at the soil surface an estimate of saturated hydraulic conductivity is produced, with all pores filled with water. As the potential is decreased progressively larger cracks drain. By measuring at the hydraulic conductivity at a range of negative potentials, the pore size distribution can be determined. While this is not a physical model of the cracks, it does give an indication to the sizes of pores within the soil.

Horgan and Young Model

Horgan and Young (2000) produced a computer model to create a two-dimensional prediction of surface crack formation. It used the fact that once cracks come within a certain distance of one another they tend to be attracted to each other. Cracks also tend to turn within a particular range of angles and at some stage a surface aggregate gets to a size that no more cracking will occur. These are often characteristic of a soil and can therefore be measured in the field and used in the model. However it was not able to predict the points at which cracking starts and although random in the formation of crack pattern, in many ways, cracking of soil is often not random, but follows lines of weaknesses.

Araldite-Impregnation Imaging

A large core sample is collected. This is then impregnated with araldite and a fluorescent resin. The core is then cut back using a grinding implement, very gradually (~1 mm per time), and at every interval the surface of the core sample is digitally imaged. The images are then loaded into a computer where they can be analysed. Depth, continuity, surface area and a number of other measurements can then be made on the cracks within the soil.

Electrical Resistivity Imaging

Using the infinite resistivity of air, the air spaces within a soil can be mapped. A specially designed resistivity meter had improved the meter-soil contact and therefore the area of the reading. This technology can be used to produce images that can be analysed for a range of cracking properties.

Soil Color

Soil color does not affect the behavior and use of soil; however, it can indicate the composition of the soil and give clues to the conditions that the soil is subjected to. Soil can exhibit a wide range of colour; gray, black, white, reds, browns, yellows and under the right conditions green. Varying horizontal bands of colour in the soil often identify a specific soil horizon. The development and distribution of color in soil results from chemical and biological weathering, especially redox reactions. As the primary minerals in soil parent material weather, the elements combine into new and colorful compounds. Aerobic conditions produce uniform or gradual color changes, while reducing environments result in disrupted color flow with complex, mottled patterns and points of color concentration.

Causes

Soil color is influenced by the amount of proteins present in the soil. Yellow or red soil indicates the presence of iron oxides. Dark brown or black color in soil indicates that the soil has a high organic matter content. Wet soil will appear darker than dry soil. However, the presence of water also affects soil color by affecting the oxidation rate. Soil that has a high water content will have less air in the soil, specifically less oxygen. In well drained (and therefore oxygen rich soils) red and brown colors caused by oxidation are more common, as opposed to in wet (low oxygen) soils where the soil usually appears grey.

The presence of specific minerals can also affect soil color. Manganese oxide causes a black color, glauconite makes the soil green, and calcite can make soil in arid regions appear white.

Classification

Often described by using general terms, such as dark brown, yellowish brown, etc., soil colors are also described more technically by using Munsell soil color charts, which separate color into components of hue (relation to red, yellow and blue), value (lightness or darkness) and chroma (paleness or strength).

Soil Water (Retention)

Soils can process and hold considerable amount of water. They can take in water, and will keep doing so until they are full, or until the rate at which they can transmit water into and through the pores is exceeded. Some of this water will steadily drain through the soil (via gravity) and end up in the waterways and streams, but much of it will be retained, despite the influence of gravity. Much of this retained water can be used by plants and other organisms, thus contributing to land productivity and soil health.

Soil Water Retention

Pores (the spaces that exist between soil particles) provide for the passage and/or retention of gasses and moisture within the soil profile. The soil's ability to retain water is strongly related to particle size; water molecules hold more tightly to the fine particles of a clay soil than to coarser particles of a sandy soil, so clays generally retain more water. Conversely, sands provide easier passage or transmission of water through the profile. Clay type, organic content, and soil structure also influence soil water retention.

The maximum amount of water that a given soil can retain is called field capacity, whereas a soil so dry that plants cannot liberate the remaining moisture from the soil particles is said to be at wilting point. Available water is that which the plants can utilize from the soil within the range between field capacity and wilting point. Roughly speaking for agriculture (top layer soil), soil is 25% water, 25% air, 45% mineral, 5% other; water varies widely from about 1% to 90% due to several retention and drainage properties of a given soil.

The role of soil water retention is profound; its effects are far reaching and relationships are invariably complex. This section focuses on a few key roles and recognizes that it is beyond the scope of this discussion to encompass all roles that can be found in the literature.

The process by which soil absorbs water and water drains downwards is called percolation.

Soil Water Retention and Organism

Soil water retention is essential to life. It provides an ongoing supply of water to plants between periods of replenishment (infiltration), so as to allow their continued growth and survival. For example, over much of temperate Victoria, Australia, this effect is seasonal and even inter-annual; the retained soil water that has accumulated in preceding wet winters permits survival of most perennial plants over typically dry summers when monthly evaporation exceeds rainfall. Soils generally contain more nutrients, moisture, and humus.

Soil Water Retention and Climate

Soil moisture has an effect on the thermal properties of a soil profile, including conductance and heat capacity. The association of soil moisture and soil thermal properties has a significant effect on temperature-related biological triggers, including seed germination, flowering, and faunal activity. (more water causes soil to more slowly gain or lose temperature given equal heating; water has roughly double the Heat capacity of soil)

Recent climate modelling by Timbal et al. (2002) suggests a strong linkage between soil moisture and the persistence and variability of surface temperature and precipitation; further, that soil moisture is a significant consideration for the accuracy of "inter-annular" predications regarding the Australian climate.

Soil Water Retention, Water Balance, and Other Influences

The role of soil in retaining water is significant in terms of the hydrological cycle; including the relative ability of soil to hold moisture and changes in soil moisture over time:

- Soil water that is not retained or used by plants may continue downward through the profile and contribute to the water table (the permanently saturated zone at the base of the profile); this is termed "recharge". Soil that is at field capacity (among other reasons) may preclude infiltration so to increase overland flow. Both effects are associated with ground and surface water supplies, erosion, and salinity.
- Soil water can affect the structural integrity or coherence of a soil; saturated soils can become unstable and result in structural failure and mass movement. Soil water, its changes over time and management are of interest to geo-technicians and soil conservationists with an interest in maintaining soil stability.

Conclusion

Soil water retention is widely studied and reported on in the literature due to its extensive and profound role; it is a key part of the hydrological cycle, provides support to organisms, interacts with climate, and is a major consideration in ground and surface water supply, environmental and geo-technical aspects.

Soil Resistivity

Soil resistivity is a measure of how much the soil resists the flow of electricity. It is a critical factor in design of systems that rely on passing current through the Earth's surface. An understanding of the soil resistivity and how it varies with depth in the soil is necessary to design the grounding system in an electrical substation, or for lightning conductors. It is needed for design of grounding (earthing) electrodes for substations and High-voltage direct current transmission systems. It can also be a useful measure in agriculture as a proxy measurement for moisture content.

In most substations the earth is used to conduct fault current when there are ground faults on the system. In single wire earth return power transmission systems, the earth itself is used as the path of conduction from the end customers (the power consumers) back to the transmission facility. In general there is some value above which the impedance of the earth connection must not rise, and some maximum step voltage which must not be exceeded to avoid endangering people and livestock.

The soil resistivity value is subject to great variation, due to moisture, temperature and chemical content. Typical values are:

- Usual values: from 10 up to 1000 (Ω-m)
- Exceptional values: from 1000 up to 10000 (Ω-m)

The SI unit of resistivity is the Ohm-meter (Ω-m); in the United States the Ohm-centimeter (Ω-cm) is often used instead. One Ω-m is 100 Ω-cm. Sometimes the conductivity, the reciprocal of the resistivity, is quoted instead.

A wide range of typical soil resistivity values can be found in literature. Military Handbook 419 (MIL-HDBK-419A) contains reference tables and formulae for the resistance of various patterns of rods and wires buried in soil of known resistivity. Being copyright free, these numbers are widely copied, sometimes without acknowledgement.

Measurement

Because soil quality may vary greatly with depth and over a wide lateral area, estimation of soil resistivity based on soil classification provide only a rough approximation. Actual resistivity measurements are required to fully qualify the resistivity and its effects on the overall transmission system.

Several methods of resistivity measurement are frequently employed:

Wenner Method

$$\rho_E = \frac{4\cdot\pi\cdot a\cdot R_W}{1+\frac{2\cdot a}{\sqrt{a^2+4\cdot b^2}}-\frac{a}{\sqrt{a^2+b^2}}} \quad \rho_E = \frac{4\cdot\pi\cdot a\cdot R_W}{1+\frac{2\cdot a}{\sqrt{a^2+4\cdot b^2}}-\frac{a}{\sqrt{a^2+b^2}}}$$

where

ρ_E = measured apparent soil resistivity (Ωm)

a = electrode spacing (m)

b = depth of the electrodes (m)

R_W = Wenner resistance measured as "V/I" in Figure (Ω) If b is small compared to a, as is the case of probes penetrating the ground only for a short distance (as normally happens), the previous equation can be reduced to:

Schlumberger Method

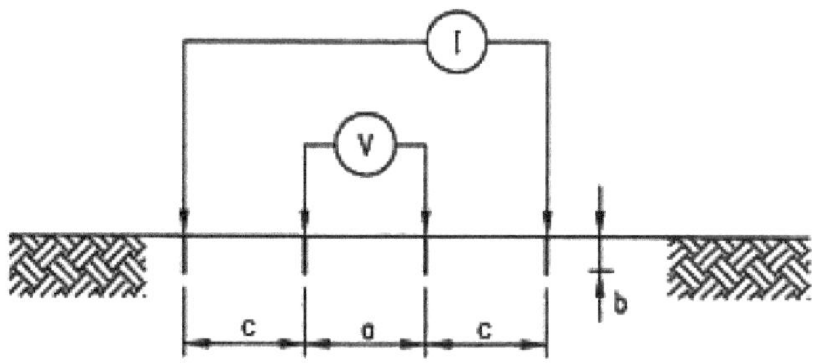

In the Schlumberger method the distance between the voltages probe is *a* and the distances from voltages probe and currents probe are *c* (see figure above).

Using the Schlumberger method, if *b* is small compared to *a* and *c*, and *c*>2*a*, the apparent soil resistivity value is:

where

ρ_E = measured apparent soil resistance (Ωm)

a = electrode spacing (m)

b = depth of the electrodes (m)

c = electrode spacing (m)

R_S = Schlumberger resistance measured as "V/I" in Figure (Ω)

Conversion Between Measures Using Schlumberger and Wenner

The conversion between values measured using the Schlumberger and Wenner methods is possible only in an approximate way. In any cases, for both Wenner and Schlumberger methods the electrode spacing between the currents probe corresponds to the depth of soil investigation and the measured apparent soil resistivity is referred to a soil volume as in the figure.

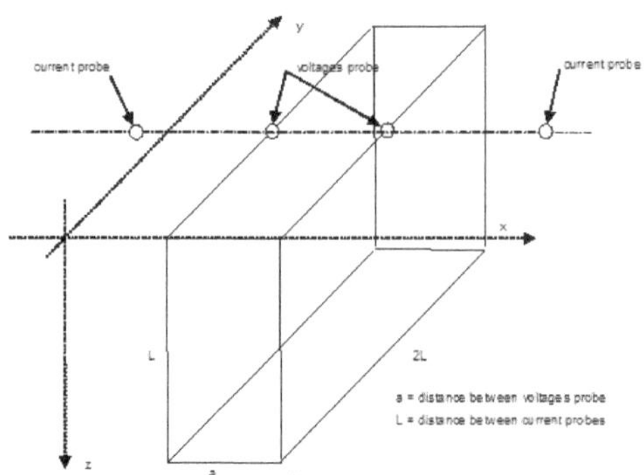

Measured volume

The current tends to flow near the surface for small probe spacing, whereas more current penetrates deeper into the soil for large spacing. The resistivity measured for a given current probe spacing represents, to a first approximation, the apparent resistivity of the soil to a depth equal to that spacing.

If the apparent soil resistivity measured with Schlumberger method ρ_E (with the corresponding electrode spacing a_s and c) is given, assuming that the soil resistivity refers to a volume as in the figure with $a=L/3$ follows:

with

$$a_W = \frac{a_S + 2c}{3}$$

where:

R_W = equivalent Wenner resistance (Ω)

a_W = equivalent electrode spacing with Wenner method (m)

a_S = electrode spacing between voltages probe with Schlumberger method (m)

c = electrode spacing between voltages and currents probe with Schlumberger method (m)

If the measured Schlumberger resistance is given, before calculating the apparent soil resistivity the following factor must be calculated:

The Wenner method is the most widely used method for measuring soil resistivity for electrical grounding (earthing) purposes. The Schlumberger method was developed to increase the voltage signal for the earlier, less sensitive instruments, by placing the potential probes closer to the current probes.

The soil resistivity measurements will be affected by existing nearby grounded electrodes. Buried conductive objects in contact with the soil can invalidate readings made by the methods described if they are close enough to alter the test current flow pattern. This is particularly true for large or long objects.

Soil Resistivity Variability

Electrical conduction in soil is essentially electrolytic and for this reason the soil resistivity depends on:

- moisture content
- salt content
- temperature (above the freezing point 0 °C)

Because of the variability of soil resistivity, IEC standards require that the seasonal variation in resistivity be accounted for in transmission system design. For practical purposes, when there is no other available information, in very cold regions, during the winter, a resistivity scale factor of 5 to 6 times the "summer" value would be an adequate average.

Corrosion

Soil resistivity is one of the driving factors determining the corrosiveness of soil. The soil corrosiveness is classified based on soil electrical resistivity by the British Standard BS-1377 as follow:

- $\rho_E > 100$ Ωm: slightly corrosive
- $50 < \rho_E < 100$ Ωm: moderately corrosive
- $10 < \rho_E < 50$ Ωm: corrosive
- $\rho_E < 10$ Ωm: severe

References

- Mullins, CE (1997). "Hardsetting". In R Lal, WH Blum, C Valentin, BA Stewart. Methods for assessment of soil degradation. Boca Raton, FL: CRC Press. p. 121. ISBN 9780849374432. Retrieved 18 August 2016.
- Brewer, Roy (1964). Fabric and mineral analysis of soils. Huntington, N.Y.: R.E. Krieger. ISBN 0882753142.
- Chesworth, Ward (2008). Encyclopedia of soil science. Dordrecht, Netherlands: Springer. p. 694. ISBN 1402039948. Retrieved 2 July 2016.
- Soil Science Glossary Terms Committee (2008). Glossary of Soil Science Terms 2008. Madison, WI: Soil Science Society of America. ISBN 978-0-89118-851-3.
- Malcolm E. Sumner (31 August 1999). Handbook of Soil Science. CRC Press. p. A-232. ISBN 978-0-8493-3136-7.
- Soil Survey Division Staff (1993). Soil survey manual. United States Department of Agriculture. pp. 63–65. Retrieved 30 August 2014.

4

Essential Concepts of Soil Science

The study of soil as a natural resource is soil science. Geologists, physical geographers, ecologists, microbiologists and specialists in regional planning all contribute to the further knowledge of soil and the advancement of the soil sciences. The essential concepts of soil science discussed in this chapter are soil compaction, soil health, soil pH, nitrogen fixation and tilth. The chapter strategically encompasses and incorporates the major components and key concepts of soil science, providing a complete understanding.

Soil Compaction

In geotechnical engineering, soil compaction is the process in which a stress applied to a soil causes densification as air is displaced from the pores between the soil grains. When stress is applied that causes densification due to water (or other liquid) being displaced from between the soil grains, then consolidation, not compaction, has occurred. Normally, compaction is the result of heavy machinery compressing the soil, but it can also occur due to the passage of (e.g.) animal feet.

A crawler-backhoe is here equipped with a narrow sheepsfoot roller to compact the fill over newly placed sewer pipe, forming a stable support for a new road surface.

In soil science and agronomy, soil compaction is usually a combination of both engineering compaction and consolidation, so may occur due to a lack of water in the soil, the applied stress being internal suction due to water evaporation as well as due to passage of animal feet. Affected soils become less able to absorb rainfall, thus increasing runoff and erosion. Plants have difficulty in compacted soil because the mineral grains are pressed together, leaving little space for air and water, which are essential for root growth. Burrowing animals also find it a hostile environment, because the denser soil is more difficult to penetrate. The ability of a soil to recover from this type of compaction depends on climate, mineralogy and fauna. Soils with high shrink-swell capacity, such as vertisols, recover quickly from compaction where moisture conditions are variable (dry spells shrink the soil, causing it to crack). But clays which do not crack as they dry cannot recover from compaction on their own unless they host ground-dwelling animals such as earthworms — the Cecil soil series is an example.

A compactor/roller fitted with a sheepsfoot drum, operated by U.S. Navy Seabees.

A Hamm vibrating roller with plain drum as used for compacting asphalt and granular soils.

Wacker Neuson vibratory rammer BS 60-2i in action.

In Construction

Soil compaction is a vital part of the construction process. It is used for support of structural entities such as building foundations, roadways, walkways, and earth retaining structures to name a few. For a given soil type certain properties may deem it more or less desirable to perform adequately for a particular circumstance. In general, the preselected soil should have adequate strength, be relatively incompressible so that future settlement is not significant, be stable against volume change as water content or other factors vary, be durable and safe against deterioration, and possess proper permeability.

When an area is to be filled or backfilled the soil is placed in layers called lifts. The ability of the first fill layers to be properly compacted will depend on the condition of the natural material being covered. If unsuitable material is left in place and backfilled, it may compress over a long period under the weight of the earth fill, causing settlement cracks in the fill or in any structure supported by the fill. In order to determine if the natural soil will support the first fill layers, an area can be proofrolled. Proofrolling consists of utilizing a piece heavy construction equipment (typically, heavy compaction equipment or hauling equipment) to roll across the fill site and watching for deflections to be revealed. These areas will be indicated by the development of rutting, pumping, or ground weaving.

To ensure adequate soil compaction is achieved, project specifications will indicate the required soil density or degree of compaction that must be achieved. These specifications are generally recommended by a geotechnical engineer in a geotechnical engineering report.

The soil type - that is, grain-size distributions, shape of the soil grains, specific gravity of soil solids, and amount and type of clay minerals, present - has a great influence on the maximum dry unit weight and optimum moisture content. It also has a great influence on how the materials should be compacted in given situations. Compaction is accomplished by use of heavy equipment. In sands and gravels, the equipment usually vibrates, to cause re-orientation of the soil particles into a denser configuration. In silts and clays, a sheepsfoot roller is frequently used, to create small zones of intense shearing, which drives air out of the soil.

Determination of adequate compaction is done by determining the in-situ density of the soil and comparing it to the maximum density determined by a laboratory test. The most commonly used laboratory test is called the Proctor compaction test and there are two different methods in obtaining the maximum density. They are the standard Proctor and modified Proctor tests; the modified Proctor is more commonly used. For small dams, the standard Proctor may still be the reference.

While soil under structures and pavements needs to be compacted, it is important after construction to decompact areas to be landscaped so that vegetation can grow.

Compaction Methods

There are several means of achieving compaction of a material. Some are more appropriate for soil compaction than others, while some techniques are only suitable for particular soils or soils in particular conditions. Some are more suited to compaction of non-soil materials such as asphalt. Generally, those that can apply significant amounts of shear as well as compressive stress, are most effective.

The available techniques can be classified as:

1. Static - a large stress is slowly applied to the soil and then released.
2. Impact - the stress is applied by dropping a large mass onto the surface of the soil.
3. Vibrating - a stress is applied repeatedly and rapidly via a mechanically driven plate or hammer. Often combined with rolling compaction.
4. Gyrating - a static stress is applied and maintained in one direction while the soil is a subjected to a gyratory motion about the axis of static loading. Limited to laboratory applications.
5. Rolling - a heavy cylinder is rolled over the surface of the soil. Commonly used on sports pitches. Roller-compactors are often fitted with vibratory devices to enhance their effectiveness.
6. Kneading - shear is applied by alternating movement in adjacent positions. An example, combined with rolling compaction, is the 'sheepsfoot' roller used in waste compaction at landfills.

The construction plant available to achieve compaction is extremely varied and is described elsewhere.

Test Methods in Laboratory

Soil compactors are used to perform test methods which cover laboratory compaction methods used to determine the relationship between molding water content and dry unit weight of soils. Soil placed as engineering fill is compacted to a dense state to obtain satisfactory engineering properties such as, shear strength, compressibility, or permeability. In addition, foundation soils are often compacted to improve their engineering properties. Laboratory compaction tests provide the basis for determining the percent compaction and molding water content needed to achieve the required engineering properties, and for controlling construction to assure that the required compaction and water contents are achieved. Test methods such as EN 13286-2, EN 13286-47, ASTM D698, ASTM D1557, AASHTO T99, AASHTO T180, AASHTO T193, BS 1377:4 provide soil compaction testing procedures.

Infiltration (Hydrology)

Infiltration is the process by which water on the ground surface enters the soil. *Infiltration rate* in soil science is a measure of the rate at which soil is able to absorb rainfall or irrigation. It is measured in inches per hour or millimeters per hour. The rate decreases as the soil becomes saturated. If the precipitation rate exceeds the infiltration rate, runoff will usually occur unless there is some physical barrier. It is related to the saturated hydraulic conductivity of the near-surface soil. The rate of infiltration can be measured using an infiltrometer.

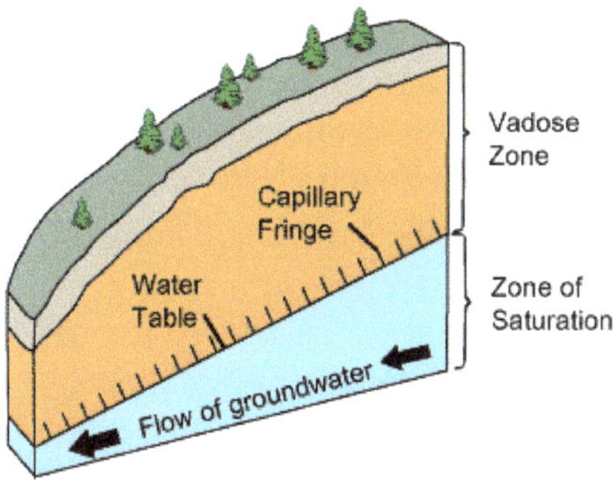

Cross-section of a hillslope depicting the vadose zone, capillary fringe, water table, and phreatic or saturated zone. *(Source: United States Geological Survey.)*

Introduction

Infiltration is caused by two forces: gravity and capillary action. While smaller pores offer greater resistance to gravity, very small pores pull water through capillary action in addition to and even against the force of gravity.

The rate of infiltration is determined by soil characteristics including ease of entry, storage capacity, and transmission rate through the soil. The soil texture and structure, vegetation types and cover, water content of the soil, soil temperature, and rainfall intensity all play a role in controlling infiltration rate and capacity. For example, coarse-grained sandy soils have large spaces between each grain and allow water to infiltrate quickly. Vegetation creates more porous soils by both protecting the soil from raindrop impact, which can close natural gaps between soil particles, and loosening soil through root action. This is why forested areas have the highest infiltration rates of any vegetative types.

The top layer of leaf litter that is not decomposed protects the soil from the pounding action of rain; without this the soil can become far less permeable. In chaparral vegetated areas, the hydrophobic oils in the succulent leaves can be spread over the soil surface with fire, creating large areas of hydrophobic soil. Other conditions that can lower infiltration rates or block them include dry plant litter that resists re-wetting, or frost.

If soil is saturated at the time of an intense freezing period, the soil can become a concrete frost on which almost no infiltration would occur. Over an entire watershed, there are likely to be gaps in the concrete frost or hygroscopic soil where water can infiltrate.

Once water has infiltrated the soil it remains in the soil, percolates down to the ground water table, or becomes part of the subsurface runoff process.

Process

The process of infiltration can continue only if there is room available for additional water at the soil surface. The available volume for additional water in the soil depends on the porosity of the soil and the rate at which previously infiltrated water can move away from the surface through the soil. The maximum rate that water can enter a soil in a given condition is the infiltration capacity. If the arrival of the water at the soil surface is less than the infiltration capacity, it is sometimes analyzed using hydrology transport models, mathematical models that consider infiltration, runoff and channel flow to predict river flow rates and stream water quality.

Research Findings

Robert E. Horton suggested that infiltration capacity rapidly declines during the early part of a storm and then tends towards an approximately constant value after a couple of hours for the remainder of the event. Previously infiltrated water fills the available storage spaces and reduces the capillary forces drawing water into the pores. Clay particles in the soil may swell as they become wet and thereby reduce the size of the pores. In areas where the ground is not protected by a layer of forest litter, raindrops can detach soil particles from the surface and wash fine particles into surface pores where they can impede the infiltration process.

Infiltration in Wastewater Collection

Wastewater collection systems consist of a set of lines, junctions and lift stations to convey sewage to a wastewater treatment plant. When these Herr lines are compromised by rupture, cracking or tree root invasion, infiltration/inflow of stormwater often occurs. This circumstance can lead to a sanitary sewer overflow, or discharge of untreated sewage to the environment.

Infiltration Calculation Methods

Infiltration is a component of the general mass balance hydrologic budget. There are several ways to estimate the volume and/or the rate of infiltration of water into a soil. The rigorous standard that fully couples groundwater to surface water through a non-homogeneous soil is the numerical solution of Richards' equation. A newer method that allows full groundwater and surface water coupling in homogeneous soil layers, and that is related to the Richards equation is the Finite water-content vadose zone flow method. In the case of uniform initial soil water content and a deep well-drained soil, there are some excellent approximate methods to solve for the infiltration flux for a single rainfall event. Among these are the Green and Ampt (1911) method, Parlange et al. (1982). Beyond these methods there are a host of empirical methods such as, SCS method, Horton's method, etc., that are little more than curve fitting exercises.

General Hydrologic Budget

The general hydrologic budget, with all the components, with respect to infiltration *F*. Given all the other variables and infiltration is the only unknown, simple algebra solves the infiltration question.

$$F = B_I + P - E - T - ET - S - I_A - R - B_O$$

where

> *F* is infiltration, which can be measured as a volume or length;
>
> B_I is the boundary input, which is essentially the output watershed from adjacent, directly connected impervious areas;
>
> B_O is the boundary output, which is also related to surface runoff, *R*, depending on where one chooses to define the exit point or points for the boundary output;
>
> *P* is precipitation;
>
> *E* is evaporation;
>
> *T* is transpiration;
>
> *ET* is evapotranspiration;
>
> *S* is the storage through either retention or detention areas;
>
> I_A is the initial abstraction, which is the short term surface storage such as puddles or even possibly detention ponds depending on size;
>
> *R* is surface runoff.

The only note on this method is one must be wise about which variables to use and which to omit, for doubles can easily be encountered. An easy example of double counting variables is when the evaporation, *E*, and the transpiration, *T*, are placed in the equation as well as the evapotranspiration, *ET*. *ET* has included in it *T* as well as a portion of *E*. Interception also needs to be accounted for, not just raw precipitation.

Richards' Equation (1931)

The standard rigorous approach for calculating infiltration into soils is Richards' Equation, which is a partial differential equation with very nonlinear coefficients. The Richards equation is computationally expensive, not guaranteed to converge, and sometimes has difficulty with mass conservation.

Finite Water-Content Vadose Zone Flow Method

This method is an approximation of the Richards' (1931) partial differential equation that de-emphasized soil water diffusivity and emphasizes advection. This approximation does not affect the calculated infiltration flux because the diffusive flux has a mean of 0. The Finite water-content vadose zone flow method is a set of three ordinary differential equations, is guaranteed to converge and to conserve mass. It requires the assumption that soil is uniform within layers.

Green and Ampt

Named for two men; Green and Ampt. The Green-Ampt method of infiltration estimation accounts for many variables that other methods, such as Darcy's law, do not. It is a function of the soil suction head, porosity, hydraulic conductivity and time.

$$\int_0^{F(t)} \frac{F}{F+\psi\Delta\theta} dF = \int_0^t K\,dt$$

where

ψ is wetting front soil suction head (L);

θ is water content (-);

K is Hydraulic conductivity (L/T);

$F(t)$ is the cumulative depth of infiltration (L).

Once integrated, one can easily choose to solve for either volume of infiltration or instantaneous infiltration rate:

$$F(t) = Kt + \psi\Delta\theta \ln\left[1+\frac{F(t)}{\psi\Delta\theta}\right].$$

Using this model one can find the volume easily by solving for $F(t)$. However the variable being solved for is in the equation itself so when solving for this one must set the variable in question to converge on zero, or another appropriate constant. A good first guess for F is the larger value of either Kt or $\sqrt{2\psi\Delta\theta Kt}$. The only note on using this formula is that one must assume that h_0, the water head or the depth of ponded water above the surface, is negligible. Using the infiltration volume from this equation one may then substitute F into the corresponding infiltration rate equation below to find the instantaneous infiltration rate at the time, t, F was measured.

$$f(t) = K\left[\frac{\psi \Delta \theta}{F(t)} + 1\right].$$

Horton's Equation

Named after the same Robert E. Horton mentioned above, Horton's equation is another viable option when measuring ground infiltration rates or volumes. It is an empirical formula that says that infiltration starts at a constant rate, f_0, and is decreasing exponentially with time, t. After some time when the soil saturation level reaches a certain value, the rate of infiltration will level off to the rate f_c.

$$f_t = f_c + (f_0 - f_c)e^{-kt}$$

Where

f_t is the infiltration rate at time t;

f_0 is the initial infiltration rate or maximum infiltration rate;

f_c is the constant or equilibrium infiltration rate after the soil has been saturated or minimum infiltration rate;

k is the decay constant specific to the soil.

The other method of using Horton's equation is as below. It can be used to find the total volume of infiltration, F, after time t.

$$F_t = f_c t + \frac{(f_0 - f_c)}{k}(1 - e^{-kt})$$

Kostiakov Equation

Named after its founder Kostiakov is an empirical equation which assumes that the intake rate declines over time according to a power function.

$$f(t) = akt^{a-1}$$

Where a and k are empirical parameters.

The major limitation of this expression is its reliance on the zero final intake rate. In most cases the infiltration rate instead approaches a finite steady value, which in some cases may occur after short periods of time. The Kostiakov-Lewis variant, also known as the "Modified Kostiakov" equation corrects for this by adding a steady intake term to the original equation.

$$f(t) = akt^{a-1} + f_0$$

in integrated form the cumulative volume is expressed as:

$$F(t) = kt^a + f_0 t$$

Where

f_0 approximates, but does not necessarily equate to the final infiltration rate of the soil.

Darcy's Law

This method used for infiltration is using a simplified version of Darcy's law. Many would argue that this method is too simple and should not be used. Compare it with the Green and Ampt (1911) solution mentioned previously. This method is similar to Green and Ampt, but missing the cumulative infiltration depth and is therefore incomplete because it assumes that the infiltration gradient occurs over some arbitrary length L. In this model the ponded water is assumed to be equal to h_0 and the head of dry soil that exists below the depth of the wetting front soil suction head is assumed to be equal to – ψ – $-\psi - L$.

$$f = K\left[\frac{h_0 - (-\psi - L)}{L}\right]$$

where

ψ is wetting front soil suction head

h_0 is the depth of ponded water above the ground surface;

K is the hydraulic conductivity;

L is the vague total depth of subsurface ground in question. This vague definition explains why this method should be avoided.

Or

$$f = K\left[\frac{L + S_f + h_0}{L}\right]$$

f Infiltration rate f (mm hour$^{-1)}$)

K is the hydraulic conductivity (mm hour$^{-1)}$);

L is the vague total depth of subsurface ground in question (mm). This vague definition explains why this method should be avoided.

S_f is wetting front soil suction head $(-\psi - \psi) = (-\psi - \psi_f)$ (mm)

h_0 is the depth of ponded water above the ground surface (mm);

Human

Early hominins—particularly the australopithecines, whose brains and anatomy are in many ways more similar to ancestral non-human apes—are less often referred to as "human" than hominins of the genus *Homo*. Several of these hominins used fire, occupied much of Eurasia, and gave rise to anatomically modern *Homo sapiens* in Africa about 200,000 years ago. They began to exhibit evidence of behavioral modernity around 50,000 years ago. In several waves of migration, anatomically modern humans ventured out of Africa and populated most of the world.

An adult human male (left) and female (right) in Northern Thailand.

Modern humans (*Homo sapiens*, primarily ssp. *Homo sapiens sapiens*) are the only extant members of Hominina clade (or human clade), a branch of the taxonomical tribe Hominini belonging to the family of great apes. They are characterized by erect posture and bipedal locomotion; manual dexterity and increased tool use, compared to other animals; and a general trend toward larger, more complex brains and societies.

The spread of humans and their large and increasing population has had a profound impact on large areas of the environment and millions of native species worldwide. Advantages that explain this evolutionary success include a relatively larger brain with a particularly well-developed neocortex, prefrontal cortex and temporal lobes, which enable high levels of abstract reasoning, language, problem solving, sociality, and culture through social learning. Humans use tools to a much higher degree than any other animal, are the only extant species known to build fires and cook their food, and are the only extant species to clothe themselves and create and use numerous other technologies and arts.

Humans are uniquely adept at utilizing systems of symbolic communication (such as language and art) for self-expression and the exchange of ideas, and for organizing themselves into purposeful groups. Humans create complex social structures composed of many cooperating and competing groups, from families and kinship networks to political states. Social interactions between humans have established an extremely wide variety of values, social norms, and rituals, which together form the basis of human society. Curiosity and the human desire to understand and influence the environment and to explain and manipulate phenomena (or events) has provided the foundation for developing science, philosophy, mythology, religion, anthropology, and numerous other fields of knowledge.

Though most of human existence has been sustained by hunting and gathering in band societies, increasing numbers of human societies began to practice sedentary agriculture approximately some 10,000 years ago, domesticating plants and animals, thus allowing for the growth of civilization. These human societies subsequently expanded in size, establishing various forms of government, religion, and culture around the world, unifying people within regions to form states and empires. The rapid advancement of scientific and medical understanding in the 19th and 20th centuries led to the development of fuel-driven technologies and increased lifespans, causing the human population to rise exponentially. By February 2016, the global human population had exceeded 7.3 billion.

In common usage, the word "human" generally refers to the only extant species of the genus *Homo* — anatomically and behaviorally modern *Homo sapiens*.

In scientific terms, the meanings of "hominid" and "hominin" have changed during the recent decades with advances in the discovery and study of the fossil ancestors of modern humans. The previously clear boundary between humans and apes has blurred, resulting in now acknowledging the hominids as encompassing multiple species, and *Homo* and close relatives since the split from chimpanzees as the only hominins. There is also a distinction between *anatomically modern humans* and *Archaic Homo sapiens*, the earliest fossil members of the species.

The English adjective *human* is a Middle English loanword from Old French *humain*, ultimately from Latin *hūmānus*, the adjective form of *homō* "man." The word's use as a noun (with a plural: *humans*) dates to the 16th century. The native English term *man* can refer to the species generally (a synonym for *humanity*), and could formerly refer to specific individuals of either sex, though this latter use is now obsolete.

The species binomial *Homo sapiens* was coined by Carl Linnaeus in his 18th century work *Systema Naturae*. The generic name *Homo* is a learned 18th century derivation from Latin *homō* "man," ultimately "earthly being" (Old Latin *hemō* a cognate to Old English *guma* "man," from PIE *d*$_{h}$▯$_{he}$*mon-*, meaning "earth" or "ground"). The species-name *sapiens* means "wise" or "sapient." Note that the Latin word *homo* refers to humans of either gender, and that *sapiens* is the singular form (while there is no such word as *sapien*).

History

Evolution and Range

The genus *Homo* evolved and diverged from other hominins in Africa, after the human clade split from the chimpanzee lineage of the hominids (great apes) branch of the primates. Modern humans, defined as the species *Homo sapiens* or specifically to the single extant subspecies *Homo sapiens sapiens*, proceeded to colonize all the continents and larger islands, arriving in Eurasia 125,000–60,000 years ago, Australia around 40,000 years ago, the Americas around 15,000 years ago, and remote islands such as Hawaii, Easter Island, Madagascar, and New Zealand between the years 300 and 1280.

Evidence from Molecular Biology

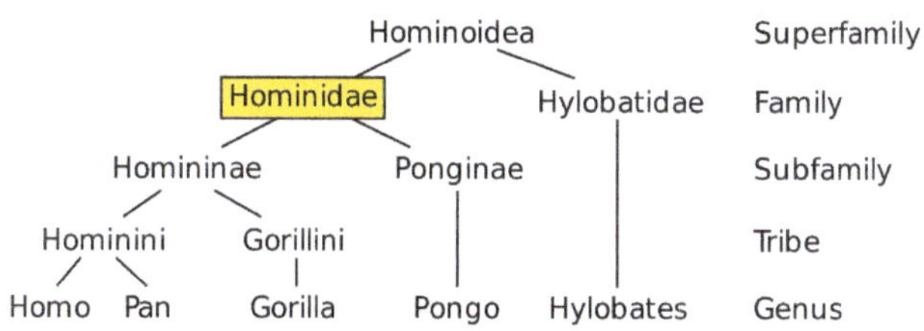

The closest living relatives of humans are chimpanzees (genus *Pan*) and gorillas (genus *Gorilla*). With the sequencing of both the human and chimpanzee genome, current estimates of similarity between human and chimpanzee DNA sequences range between 95% and 99%. By using the technique called a molecular clock which estimates the time required for the number of divergent mutations to accumulate between two lineages, the approximate date for the split between lineages can be calculated. The gibbons (*Hylobatidae*) and orangutans (genus *Pongo*) were the first groups to split from the line leading to the humans, then gorillas (genus *Gorilla*) followed by the chimpanzees (genus *Pan*). The splitting date between human and chimpanzee lineages is placed around 4–8 million years ago during the late Miocene epoch. During this split, chromosome 2 was formed from two other chromosomes, leaving humans with only 23 pairs of chromosomes, compared to 24 for the other apes.

Evidence from The Fossil Record

There is little fossil evidence for the divergence of the gorilla, chimpanzee and hominin lineages. The earliest fossils that have been proposed as members of the hominin lineage are *Sahelanthropus tchadensis* dating from 7 million years ago, *Orrorin tugenensis* dating from 5.7 million years ago, and *Ardipithecus kadabba* dating to 5.6 million years ago. Each of these species has been argued to be a bipedal ancestor of later hominins, but all such claims are contested. It is also possible that any one of the three is an ancestor of another branch of African apes, or is an ancestor shared between hominins and other African Hominoidea (apes). The question of the relation between these early fossil species and the hominin lineage is still to be resolved. From these early species the australopithecines arose around 4 million years ago diverged into robust (also called *Paranthropus*) and gracile branches, possibly one of which (such as *A. garhi*, dating to 2.5 million years ago) is a direct ancestor of the genus *Homo*.

The earliest members of the genus *Homo* are *Homo habilis* which evolved around 2.8 million years ago. *Homo habilis* has been considered the first species for which there is clear evidence of the use of stone tools. More recently, however, in 2015, stone tools, perhaps predating *Homo habilis*, have been discovered in northwestern Kenya that have been dated to 3.3 million years old. Nonetheless, the brains of *Homo habilis* were about the same size as that of a chimpanzee, and their main adaptation was bipedalism as an adaptation to terrestrial living. During the next million years a process of encephalization began, and with the arrival of *Homo erectus* in the fossil record, cranial capacity had doubled. *Homo erectus* were the first of the hominina to leave Africa, and these species spread through Africa, Asia, and Europe between 1.3 to 1.8 million years ago. One population of *H. erectus*, also sometimes classified as a separate species *Homo ergaster*, stayed in Africa and evolved into *Homo sapiens*. It is believed that these species were the first to use fire and complex tools. The earliest transitional fossils between *H. ergaster/erectus* and archaic humans are from Africa such as *Homo rhodesiensis*, but seemingly transitional forms are also found at Dmanisi, Georgia. These descendants of African *H. erectus* spread through Eurasia from ca. 500,000 years ago evolving into *H. antecessor*, *H. heidelbergensis* and *H. neanderthalensis*. The earliest fossils of anatomically modern humans are from the Middle Paleolithic, about 200,000 years ago such as the Omo remains of Ethiopia and the fossils of Herto sometimes classified as *Homo sapiens idaltu*. Later fossils of archaic *Homo sapiens* from Skhul in Israel and Southern Europe begin around 90,000 years ago.

Anatomical Adaptations

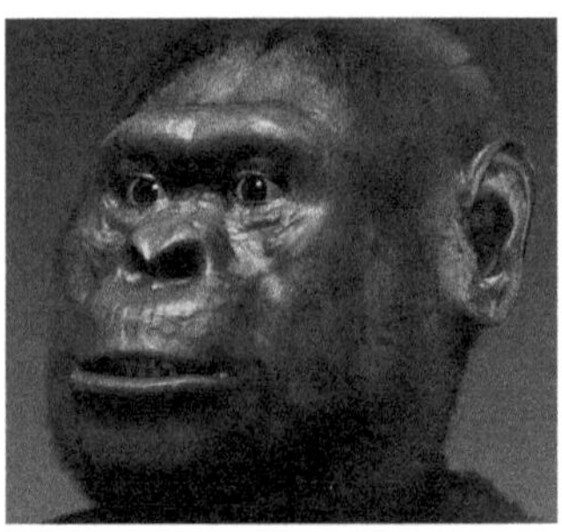

Reconstruction of *Australopithecus afarensis*.

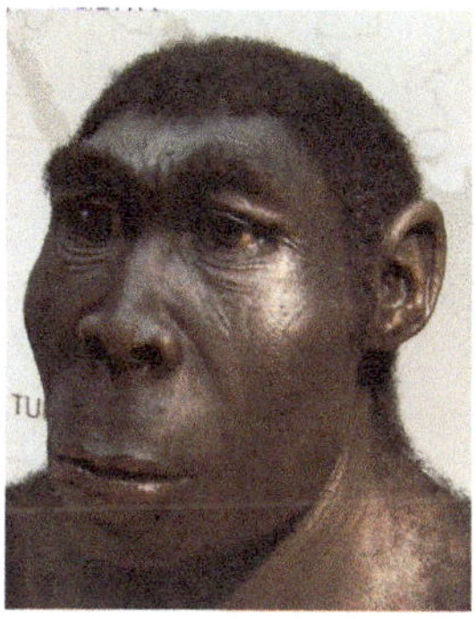

Reconstruction of *Homo erectus*

Reconstruction of *Homo habilis*, the earliest known species of the genus *Homo* and the first human ancestor to use stone tools

Reconstruction of An Adult Female Homo Erectus.

Human evolution is characterized by a number of morphological, developmental, physiological, and behavioral changes that have taken place since the split between the last common ancestor of humans and chimpanzees. The most significant of these adaptations are 1. bipedalism, 2. increased brain size, 3. lengthened ontogeny (gestation and infancy), 4. decreased sexual dimorphism (neoteny). The relationship between all these changes is the subject of ongoing debate. Other significant morphological changes included the evolution of a power and precision grip, a change first occurring in *H. erectus*.

Bipedalism is the basic adaption of the hominin line, and it is considered the main cause behind a suite of skeletal changes shared by all bipedal hominins. The earliest bipedal hominin is considered to be either *Sahelanthropus* or *Orrorin*, with *Ardipithecus*, a full bipedal, coming somewhat later. The knuckle walkers, the gorilla and chimpanzee, diverged around the same time, and either *Sahelanthropus* or *Orrorin* may be humans' last shared ancestor with those animals. The early bipedals eventually evolved into the australopithecines and later the genus *Homo*. There are several theories of the adaptational value of bipedalism. It is possible that bipedalism was favored because it freed up the hands for reaching and carrying food, because it saved energy during locomotion, because it enabled long distance running and hunting, or as a strategy for avoiding hyperthermia by reducing the surface exposed to direct sun.

The human species developed a much larger brain than that of other primates – typically 1,330 cm^3 in modern humans, over twice the size of that of a chimpanzee or gorilla. The pattern of encephalization started with *Homo habilis* which at approximately 600 cm^3 had a brain slightly larger than chimpanzees, and continued with *Homo erectus* (800–1100 cm^3), and reached a maximum in Neanderthals with an average size of 1200–1900 cm^3, larger even than *Homo sapiens* (but less encephalized). The pattern of human postnatal brain growth differs from that of other apes (heterochrony), and allows for extended periods of social learning and language acquisition in juvenile humans. However, the differences between the structure of human brains and those of other apes may be even more significant than differences in size. The increase in volume over time has affected different areas within the brain unequally – the temporal lobes, which contain centers for language processing have increased disproportionately, as has the prefrontal cortex which has been related to complex decision making and moderating social behavior. Encephalization has been tied to an increasing emphasis on meat in the diet, or with the development of cooking, and it has been proposed that intelligence increased as a response to an increased necessity for solving social problems as human society became more complex.

The reduced degree of sexual dimorphism is primarily visible in the reduction of the male canine tooth relative to other ape species (except gibbons). Another important physiological change related to sexuality in humans was the evolution of hidden estrus. Humans are the only ape in which the female is fertile year round, and in which no special signals of fertility are produced by the body (such as genital swelling during estrus). Nonetheless humans retain a degree of sexual dimorphism in the distribution of body hair and subcutaneous fat, and in the overall size, males being around 25% larger than females. These changes taken together have been interpreted as a result of an increased emphasis on pair bonding as a possible solution to the requirement for increased parental investment due to the prolonged infancy of offspring.

Rise of Homo Sapiens

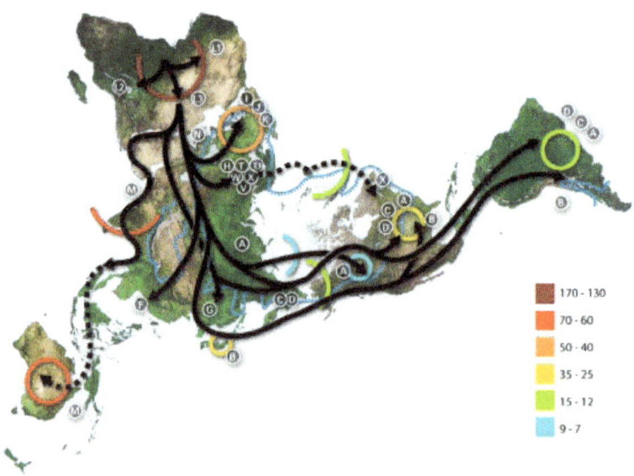

World map of early human migrations according to mitochondrial population genetics (numbers are millennia before present, the North Pole is at the center).

By the beginning of the Upper Paleolithic period (50,000 BP), full behavioral modernity, including language, music and other cultural universals had developed. As modern humans spread out from Africa they encountered other hominids such as *Homo neanderthalensis* and the so-called Denisovans. The nature of interaction between early humans and these sister species has been a long-standing source of controversy, the question being whether humans replaced these earlier species or whether they were in fact similar enough to interbreed, in which case these earlier populations may have contributed genetic material to modern humans. Recent studies of the human and Neanderthal genomes suggest gene flow between archaic *Homo sapiens* and Neanderthals and Denisovans. In March 2016, studies were published that suggest that modern humans bred with hominins, including Denisovans and Neanderthals, on multiple occasions.

This dispersal out of Africa is estimated to have begun about 70,000 years BP from Northeast Africa. Current evidence suggests that there was only one such dispersal and that it only involved a few hundred individuals. The vast majority of humans stayed in Africa and adapted to a diverse array of environments. Modern humans subsequently spread globally, replacing earlier hominins (either through competition or hybridization). They inhabited Eurasia and Oceania by 40,000 years BP, and the Americas at least 14,500 years BP.

Transition to Civilization

Until about 10,000 years ago, humans lived as hunter-gatherers. They gradually gained domination over much of the natural environment. They generally lived in small nomadic groups known as band societies, often in caves. The advent of agriculture prompted the Neolithic Revolution, when access to food surplus led to the formation of permanent human settlements, the domestication of animals and the use of metal tools for the first time in history. Agriculture encouraged trade and cooperation, and led to complex society.

The rise of agriculture, and domestication of animals, led to stable human settlements.

The early civilisations of Mesopotamia, Egypt, India, China, Maya, Greece and Rome were some of the cradles of civilisation. The Late Middle Ages and the Early Modern Period saw the rise of revolutionary ideas and technologies. Over the next 500 years, exploration and European colonialism brought great parts of the world under European control, leading to later struggles for independence. The concept of the modern world as distinct from an ancient world is based on a rapid change progress in a brief period of time in many areas. Advances in all areas of human activity prompted new theories such as evolution and psychoanalysis, which changed humanity's views of itself. The Scientific Revolution, Technological Revolution and the Industrial Revolution up until the 19th century resulted in independent discoveries such as imaging technology, major innovations in transport, such as the airplane and automobile; energy development, such as coal and electricity. This correlates with population growth (especially in America) and higher life expectancy, the World population rapidly increased numerous times in the 19th and 20th centuries as nearly 10% of the 100 billion people lived in the past century.

With the advent of the Information Age at the end of the 20th century, modern humans live in a world that has become increasingly globalized and interconnected. As of 2010, almost 2 billion humans are able to communicate with each other via the Internet, and 3.3 billion by mobile phone subscriptions. Although interconnection between humans has encouraged the growth of science, art, discussion, and technology, it has also led to culture clashes and the development and use of weapons of mass destruction. Human civilization has led to environmental destruction and pollution significantly contributing to the ongoing mass extinction of other forms of life called the Holocene extinction event, which may be further accelerated by global warming in the future.

Habitat and Population

The Earth, as seen from space in October 2000, showing the extent of human occupation of the planet. The bright lights signify both the most densely inhabited areas and ones financially capable of illuminating those areas.

Early human settlements were dependent on proximity to water and, depending on the lifestyle, other natural resources used for subsistence, such as populations of animal prey for hunting and arable land for growing crops and grazing livestock. But humans have a great capacity for altering their habitats by means of technology, through irrigation, urban planning, construction, transport, manufacturing goods, deforestation and desertification. Deliberate habitat alteration is often done with the goals of increasing material wealth, increasing thermal comfort, improving the amount of food available, improving aesthetics, or improving ease of access to resources or other human settlements. With the advent of large-scale trade and transport infrastructure, proximity to these resources has become unnecessary, and in many places, these factors are no longer a driving force behind the growth and decline of a population. Nonetheless, the manner in which a habitat is altered is often a major determinant in population change.

Technology has allowed humans to colonize all of the continents and adapt to virtually all climates. Within the last century, humans have explored Antarctica, the ocean depths, and outer space, although large-scale colonization of these environments is not yet feasible. With a population of over seven billion, humans are among the most numerous of the large mammals. Most humans (61%) live in Asia. The remainder live in the Americas (14%), Africa (14%), Europe (11%), and Oceania (0.5%).

Human habitation within closed ecological systems in hostile environments, such as Antarctica and outer space, is expensive, typically limited in duration, and restricted to scientific, military, or industrial expeditions. Life in space has been very sporadic, with no more than thirteen humans in space at any given time. Between 1969 and 1972, two humans at a time spent brief intervals on the Moon. As of September 2016, no other celestial body has been visited by humans, although there has been a continuous human presence in space since the launch of the initial crew to inhabit the International Space Station on October 31, 2000. However, other celestial bodies have been visited by human-made objects.

Since 1800, the human population has increased from one billion to over seven billion, In 2004, some 2.5 billion out of 6.3 billion people (39.7%) lived in urban areas. In February 2008, the U.N. estimated that half the world's population would live in urban areas by the end of the year. Problems for humans living in cities include various forms of pollution and crime, especially in inner city and suburban slums. Both overall population numbers and the proportion residing in cities are expected to increase significantly in the coming decades.

Humans have had a dramatic effect on the environment. Humans are apex predators, being rarely preyed upon by other species. Currently, through land development, combustion of fossil fuels, and pollution, humans are thought to be the main contributor to global climate change. If this continues at its current rate it is predicted that climate change will wipe out half of all plant and animal species over the next century.

Biology

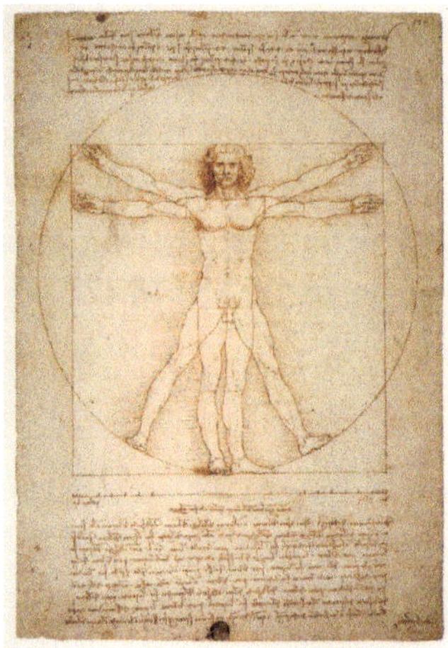

Vitruvian Man, Leonardo da Vinci's image is often used as an implied symbol of the essential symmetry of the human body, and by extension, of the universe as a whole.

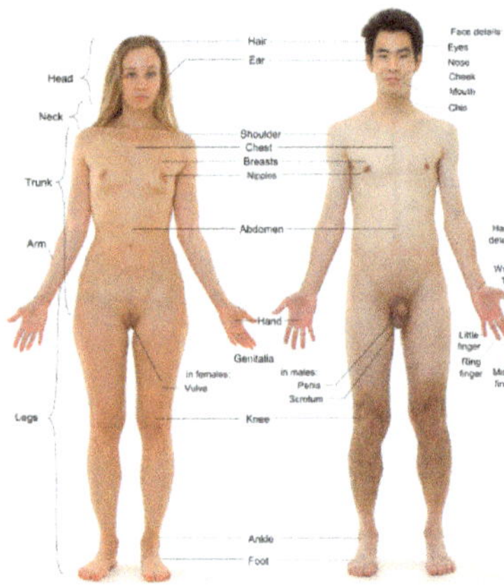

Basic anatomical features of female and male humans. These models have had body hair and male facial hair removed and head hair trimmed. The female model is wearing red nail polish on her toenails and a ring.

Anatomy and Physiology

Most aspects of human physiology are closely homologous to corresponding aspects of animal physiology. The human body consists of the legs, the torso, the arms, the neck, and the head. An adult human body consists of about 100 trillion (10^{14}) cells. The most commonly defined body systems in humans are the nervous, the cardiovascular, the circulatory, the digestive, the endocrine, the immune, the integumentary, the lymphatic, the muscoskeletal, the reproductive, the respiratory, and the urinary system.

Humans, like most of the other apes, lack external tails, have several blood type systems, have opposable thumbs, and are sexually dimorphic. The comparatively minor anatomical differences between humans and chimpanzees are a result of human bipedalism. One difference is that humans have a far faster and more accurate throw than other animals. Humans are also among the best long-distance runners in the animal kingdom, but slower over short distances. Humans' thinner body hair and more productive sweat glands help avoid heat exhaustion while running for long distances.

As a consequence of bipedalism, human females have narrower birth canals. The construction of the human pelvis differs from other primates, as do the toes. A trade-off for these advantages of the modern human pelvis is that childbirth is more difficult and dangerous than in most mammals, especially given the larger head size of human babies compared to other primates. This means that human babies must turn around as they pass through the birth canal, which other primates do not do, and it makes humans the only species where females require help from their conspecifics to reduce the risks of birthing. As a partial evolutionary solution, human fetuses are born less developed and more vulnerable. Chimpanzee babies are cognitively more developed than human babies until the age of six months, when the rapid development of human brains surpasses chimpanzees. Another difference between women and chimpanzee females is that women go through the menopause and become unfertile decades before the end of their lives. All species of non-human apes are capable of giving birth until death. Menopause probably developed as it has provided an evolutionary advantage (more caring time) to young relatives.

Apart from bipedalism, humans differ from chimpanzees mostly in smelling, hearing, digesting proteins, brain size, and the ability of language. Humans' brains are about three times bigger than in chimpanzees. More importantly, the brain to body ratio is much higher in humans than in chimpanzees, and humans have a significantly more developed cerebral cortex, with a larger number of neurons. The mental abilities of humans are remarkable compared to other apes. Humans' ability of speech is unique among primates. Humans are able to create new and complex ideas, and to develop technology, which is unprecedented among other organisms on Earth.

It is estimated that the worldwide average height for an adult human male is about 172 cm (5 ft 7 ½ in), while the worldwide average height for adult human females is about 158 cm (5 ft 2 in). Shrinkage of stature may begin in middle age in some individuals, but tends to be typical in the extremely aged. Through history human populations have universally become taller, probably as a consequence of better nutrition, healthcare, and living conditions. The average mass of an adult human is 54–64 kg (120–140 lb) for females and 76–83 kg (168–183 lb) for males. Like many other conditions, body weight and body type is influenced by both genetic susceptibility and environment and varies greatly among individuals.

Although humans appear hairless compared to other primates, with notable hair growth occurring chiefly on the top of the head, underarms and pubic area, the average human has more hair follicles on his or her body than the average chimpanzee. The main distinction is that human hairs are shorter, finer, and less heavily pigmented than the average chimpanzee's, thus making them harder to see. Humans have about 2 million sweat glands spread over their entire bodies, many more than chimpanzees, whose sweat glands are scarce and are mainly located on the palm of the hand and on the soles of the feet.

The dental formula of humans is: 2.1.2.32.1.2.3. Humans have proportionately shorter palates and much smaller teeth than other primates. They are the only primates to have short, relatively flush canine teeth. Humans have characteristically crowded teeth, with gaps from lost teeth usually closing up quickly in young individuals. Humans are gradually losing their wisdom teeth, with some individuals having them congenitally absent.

Genetics

Like all mammals, humans are a diploid eukaryotic species. Each somatic cell has two sets of 23 chromosomes, each set received from one parent; gametes have only one set of chromosomes, which is a mixture of the two parental sets. Among the 23 pairs of chromosomes there are 22 pairs of autosomes and one pair of sex chromosomes. Like other mammals, humans have an XY sex-determination system, so that females have the sex chromosomes XX and males have XY.

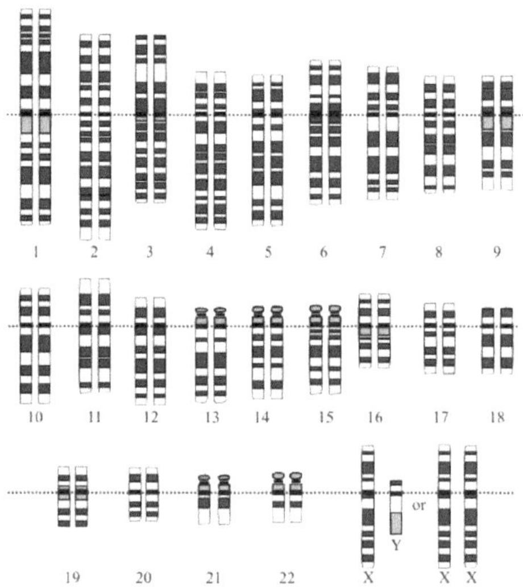

A graphical representation of the ideal human karyotype, including both the male and female variant of the sex chromosome (number 23).

One human genome was sequenced in full in 2003, and currently efforts are being made to achieve a sample of the genetic diversity of the species. By present estimates, humans have approximately 22,000 genes. The variation in human DNA is very small compared to other species, possibly suggesting a population bottleneck during the Late Pleistocene (around 100,000 years ago), in which the human population was reduced to a small number of breeding pairs. Nucleotide diversity is based on single mutations called single nucleotide polymorphisms (SNPs). The nucleotide diversity between humans is about 0.1%, i.e. 1 difference per 1,000 base pairs. A difference of 1 in 1,000 nucleotides between two humans chosen at random amounts to about 3 million nucleotide differences, since the human genome has about 3 billion nucleotides. Most of these single nucleotide polymorphisms (SNPs) are neutral but some (about 3 to 5%) are functional and influence phenotypic differences between humans through alleles.

By comparing the parts of the genome that are not under natural selection and which therefore accumulate mutations at a fairly steady rate, it is possible to reconstruct a genetic tree incorporating the entire human species since the last shared ancestor. Each time a certain mutation (SNP) appears in an individual and is passed on to his or her descendants, a haplogroup is formed including all of the descendants of the individual who will also carry that mutation. By comparing mitochondrial DNA, which is inherited only from the mother, geneticists have concluded that the last female common ancestor whose genetic marker is found in all modern humans, the so-called mitochondrial Eve, must have lived around 90,000 to 200,000 years ago.

Human accelerated regions, first described in August 2006, are a set of 49 segments of the human genome that are conserved throughout vertebrate evolution but are strikingly different in humans. They are named according to their degree of difference between humans and their nearest animal relative (chimpanzees) (HAR1 showing the largest degree of human-chimpanzee differences). Found by scanning through genomic databases of multiple species, some of these highly mutated areas may contribute to human-specific traits.

The forces of natural selection have continued to operate on human populations, with evidence that certain regions of the genome display directional selection in the past 15,000 years.

Life cycle

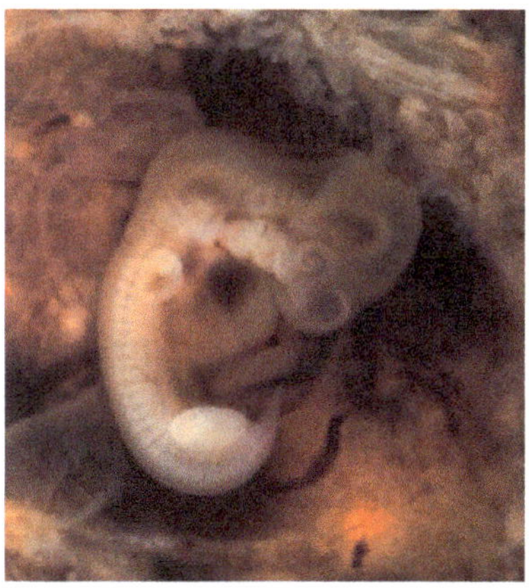

A 10 mm human embryo at 5 weeks

Boy and girl before puberty

Adult man and woman

Youth male and female

As with other mammals, human reproduction takes place as internal fertilization by sexual intercourse. During this process, the male inserts his erect penis into the female's vagina and ejaculates semen, which contains sperm. The sperm travels through the vagina and cervix into the uterus or Fallopian tubes for fertilization of the ovum. Upon fertilization and implantation, gestation then occurs within the female's uterus.

Elderly man and woman

The zygote divides inside the female's uterus to become an embryo, which over a period of 38 weeks (9 months) of gestation becomes a fetus. After this span of time, the fully grown fetus is birthed from the woman's body and breathes independently as an infant for the first time. At this point, most modern cultures recognize the baby as a person entitled to the full protection of the law, though some jurisdictions extend various levels of personhood earlier to human fetuses while they remain in the uterus.

Compared with other species, human childbirth is dangerous. Painful labors lasting 24 hours or more are not uncommon and sometimes lead to the death of the mother, the child or both. This is because of both the relatively large fetal head circumference and the mother's relatively narrow pelvis. The chances of a successful labor increased significantly during the 20th century in wealthier countries with the advent of new medical technologies. In contrast, pregnancy and natural childbirth remain hazardous ordeals in developing regions of the world, with maternal death rates approximately 100 times greater than in developed countries.

In developed countries, infants are typically 3–4 kg (6–9 pounds) in weight and 50–

60 cm (20–24 inches) in height at birth. However, low birth weight is common in developing countries, and contributes to the high levels of infant mortality in these regions. Helpless at birth, humans continue to grow for some years, typically reaching sexual maturity at 12 to 15 years of age. Females continue to develop physically until around the age of 18, whereas male development continues until around age 21. The human life span can be split into a number of stages: infancy, childhood, adolescence, young adulthood, adulthood and old age. The lengths of these stages, however, have varied across cultures and time periods. Compared to other primates, humans experience an unusually rapid growth spurt during adolescence, where the body grows 25% in size. Chimpanzees, for example, grow only 14%, with no pronounced spurt. The presence of the growth spurt is probably necessary to keep children physically small until they are psychologically mature. Humans are one of the few species in which females undergo menopause. It has been proposed that menopause increases a woman's overall reproductive success by allowing her to invest more time and resources in her existing offspring, and in turn their children (the grandmother hypothesis), rather than by continuing to bear children into old age.

For various reasons, including biological/genetic causes, women live on average about four years longer than men—as of 2013 the global average life expectancy at birth of a girl is estimated at 70.2 years compared to 66.1 for a boy. There are significant geographical variations in human life expectancy, mostly correlated with economic development—for example life expectancy at birth in Hong Kong is 84.8 years for girls and 78.9 for boys, while in Swaziland, primarily because of AIDS, it is 31.3 years for both sexes. The developed world is generally aging, with the median age around 40 years. In the developing world the median age is between 15 and 20 years. While one in five Europeans is 60 years of age or older, only one in twenty Africans is 60 years of age or older. The number of centenarians (humans of age 100 years or older) in the world was estimated by the United Nations at 210,000 in 2002. At least one person, Jeanne Calment, is known to have reached the age of 122 years; higher ages have been claimed but they are not well substantiated.

Diet

Humans living in Bali, Indonesia preparing a meal.

Humans are omnivorous, capable of consuming a wide variety of plant and animal material. Varying with available food sources in regions of habitation, and also varying with cultural and religious norms, human groups have adopted a range of diets, from purely vegetarian to primarily carnivorous. In some cases, dietary restrictions in humans can lead to deficiency diseases; however, stable human groups have adapted to many dietary patterns through both genetic specialization and cultural conventions to use nutritionally balanced food sources. The human diet is prominently reflected in human culture, and has led to the development of food science.

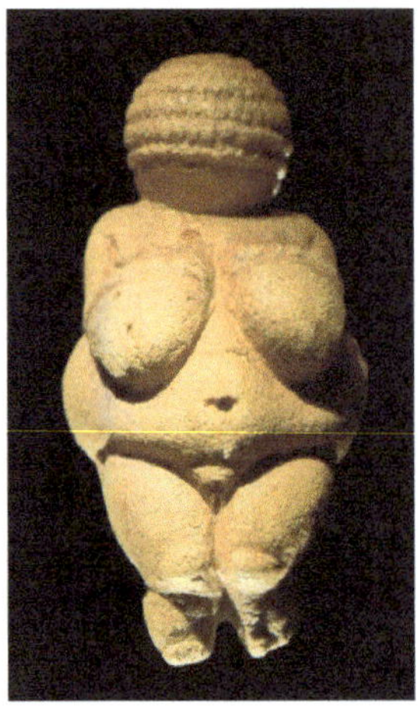

Venus of Willendorf statuette from the Upper Palaeolithic period

Until the development of agriculture approximately 10,000 years ago, *Homo sapiens* employed a hunter-gatherer method as their sole means of food collection. This involved combining stationary food sources (such as fruits, grains, tubers, and mushrooms, insect larvae and aquatic mollusks) with wild game, which must be hunted and killed in order to be consumed. It has been proposed that humans have used fire to prepare and cook food since the time of *Homo erectus*. Around ten thousand years ago, humans developed agriculture, which substantially altered their diet. This change in diet may also have altered human biology; with the spread of dairy farming providing a new and rich source of food, leading to the evolution of the ability to digest lactose in some adults. Agriculture led to increased populations, the development of cities, and because of increased population density, the wider spread of infectious diseases. The types of food consumed, and the way in which they are prepared, have varied widely by time, location, and culture.

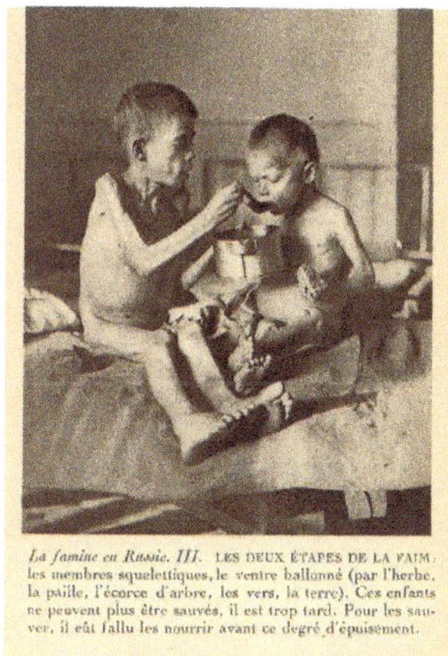

Two starved boys during the Russian famine of 1921–22

In general, humans can survive for two to eight weeks without food, depending on stored body fat. Survival without water is usually limited to three or four days. About 36 million humans die every year from causes directly or indirectly related to starvation. Childhood malnutrition is also common and contributes to the global burden of disease. However global food distribution is not even, and obesity among some human populations has increased rapidly, leading to health complications and increased mortality in some developed, and a few developing countries. Worldwide over one billion people are obese, while in the United States 35% of people are obese, leading to this being described as an "obesity epidemic." Obesity is caused by consuming more calories than are expended, so excessive weight gain is usually caused by an energy-dense diet.

Biological Variation

People in warm climates are often slender and dark skinned, such as these Maasai men from Kenya.

No two humans—not even monozygotic twins—are genetically identical. Genes and environment influence human biological variation from visible characteristics to physiology to disease susceptibly to mental abilities. The exact influence of genes and environment on certain traits is not well understood.

People in cold climates tend to be lighter skinned such as these Inuit women from Canada.

Most current genetic and archaeological evidence supports a recent single origin of modern humans in East Africa, with first migrations placed at 60,000 years ago. Compared to the great apes, human gene sequences—even among African populations—are remarkably homogeneous. On average, genetic similarity between any two humans is 99.9%. There is about 2–3 times more genetic diversity within the wild chimpanzee population, than in the entire human gene pool.

The human body's ability to adapt to different environmental stresses is remarkable, allowing humans to acclimatize to a wide variety of temperatures, humidity, and altitudes. As a result, humans are a cosmopolitan species found in almost all regions of the world, including tropical rainforests, arid desert, extremely cold arctic regions, and heavily polluted cities. Most other species are confined to a few geographical areas by their limited adaptability.

There is biological variation in the human species—with traits such as blood type, cranial features, eye color, hair color and type, height and build, and skin color varying across the globe. Human body types vary substantially. The typical height of an adult human is between 1.4 m and 1.9 m (4 ft 7 in and 6 ft 3 in), although this varies significantly depending, among other things, on sex and ethnic origin. Body size is partly determined by genes and is also significantly influenced by environmental factors such as diet, exercise, and sleep patterns, especially as an influence in childhood. Adult height for each sex in a particular ethnic group approximately follows a normal distribution. Those aspects of genetic variation that give clues to human evolutionary history, or are relevant to medical research, have received particular attention. For example, the genes that allow adult humans to digest lactose are present in high frequencies in populations that have long histories of cattle domestication, suggesting natural selection having favored that gene in populations that depend on cow milk. Some hereditary diseases such as sickle cell anemia are frequent in populations where malaria has been endemic throughout history—it is believed that the same gene gives increased resistance to malaria among those who are unaffected carriers of the gene. Similarly, populations that have for a long time inhabited specific climates, such as arctic or tropical regions or high altitudes, tend to have developed specific phenotypes that are beneficial for conserving energy in those environments—short stature and stocky build in cold regions, tall and lanky in hot regions, and with high lung capacities at high altitudes. Similarly, skin color varies clinally with darker skin around the equator—where the added protection from the sun's ultraviolet radiation is thought to give an evolutionary advantage—and lighter skin tones closer to the poles.

The hue of human skin and hair is determined by the presence of pigments called melanins. Human skin color can range from darkest brown to lightest peach, or even nearly white or colorless in cases of albinism. Human hair ranges in color from white to red to blond to brown to black, which is most frequent. Hair color depends on the amount of melanin (an effective sun blocking pigment) in the skin and hair, with hair melanin concentrations in hair fading with increased age, leading to grey or even white hair. Most researchers believe that skin darkening is an adaptation that evolved as protection against ultraviolet solar radiation, which also helps balancing folate, which is destroyed by ultraviolet radiation. Light skin pigmentation protects against depletion of vitamin D, which requires sunlight to make. Skin pigmentation of contemporary humans is clinally distributed across the planet, and in general correlates with the level of ultraviolet radiation in a particular geographic area. Human skin also has a capacity to darken (tan) in response to exposure to ultraviolet radiation.

Structure of Variation

A Libyan, a Nubian, a Syrian, and an Egyptian, drawing by an unknown artist after a mural of the tomb of Seti I.

The ancestors of Native Americans, such as this Yanomami woman, crossed into the Americas from Northeast Asia, and genetic and linguistic evidence links them to North Asian populations, particularly those of East Siberia.

An older adult human male European in Paris – playing chess at the Jardins du Luxembourg.

Within the human species, the greatest degree of genetic variation exists between males and females. While the nucleotide genetic variation of individuals of the same sex across global populations is no greater than 0.1%, the genetic difference between males and females is between 1% and 2%. Although different in nature, this approaches the genetic differentiation between men and male chimpanzees or women and female chimpanzees. The genetic difference between sexes contributes to anatomical, hormonal, neural, and physiological differences between men and women, although the exact degree and nature of social and environmental influences on sexes are not completely understood. Males on average are 15% heavier and 15 cm taller than females. There is a difference between body types, body organs and systems, hormonal levels, sensory systems, and muscle mass between sexes. On average, there is a difference of about 40–50% in upper body strength and 20–30% in lower body strength between men and women. Women generally have a higher body fat percentage than men. Women have lighter skin than men of the same population; this has been explained by a higher need for vitamin D (which is synthesized by sunlight) in females during pregnancy and lactation. As there are chromosomal differences between females and males, some X and Y chromosome related conditions and disorders only affect either men or women. Other conditional differences between males and females are not related to sex chromosomes. Even after allowing for body weight and volume, the male voice is usually an octave deeper than the female voice. Women have a longer life span in almost every population around the world.

Males typically have larger tracheae and branching bronchi, with about 30% greater lung volume per unit body mass. They have larger hearts, 10% higher red blood cell count, and higher hemoglobin, hence greater oxygen-carrying capacity. They also have higher circulating clotting factors (vitamin K, prothrombin and platelets). These differences lead to faster healing of wounds and higher peripheral pain tolerance. Females typically have more white blood cells (stored and circulating), more granulocytes and B and T lymphocytes. Additionally, they produce more antibodies at a faster rate than males. Hence they develop fewer infectious diseases and these continue for shorter periods. Ethologists argue that females, interacting with other females and multiple offspring in social groups, have experienced such traits as a selective advantage. According to Daly and Wilson, "The sexes differ more in human beings than in monogamous mammals, but much less than in extremely polygamous mammals." But given that sexual dimorphism in the closest relatives of humans is much greater than among humans, the human clade must be considered to be characterized by decreasing sexual dimorphism, probably due to less competitive mating patterns. One proposed explanation is that human sexuality has developed more in common with its close relative the bonobo, which exhibits similar sexual dimorphism, is polygynandrous and uses recreational sex to reinforce social bonds and reduce aggression.

Humans of the same sex are 99.9% genetically identical. There is extremely little variation between human geographical populations, and most of the variation that does occur is at the personal level within local areas, and not between populations. Of the 0.1% of human genetic differentiation, 85% exists within any randomly chosen local population, be they Italians, Koreans, or Kurds. Two randomly chosen Koreans may be genetically as different as a Korean and an Italian. Any ethnic group contains 85% of the human genetic diversity of the world. Genetic data shows that no matter how population groups are defined, two people from the same population group are about as different from each other as two people from any two different population groups.

Current genetic research has demonstrated that humans on the African continent are the most genetically diverse. There is more human genetic diversity in Africa than anywhere else on Earth. The genetic structure of Africans was traced to 14 ancestral population clusters. Human genetic diversity decreases in native populations with migratory distance from Africa and this is thought to be the result of bottlenecks during human migration. Humans have lived in Africa for the longest time, which has allowed accumulation of a higher diversity of genetic mutations in these populations. Only part of Africa's population migrated out of the continent, bringing just part of the original African genetic variety with them. African populations harbor genetic alleles that are not found in other places of the world. All the common alleles found in populations outside of Africa are found on the African continent.

Geographical distribution of human variation is complex and constantly shifts through time which reflects complicated human evolutionary history. Most human biological variation is clinally distributed and blends gradually from one area to the next. Groups of people around the world have different frequencies of polymorphic genes. Furthermore, different traits are non-concordant and each have different clinal distribution. Adaptability varies both from person to person and from population to population. The most efficient adaptive responses are found in geographical populations where the environmental stimuli are the strongest (e.g. Tibetans are highly adapted to high altitudes). The clinal geographic genetic variation is further complicated by the migration and mixing between human populations which has been occurring since prehistoric times.

Human variation is highly non-concordant: most of the genes do not cluster together and are not inherited together. Skin and hair color are not correlated to height, weight, or athletic ability. Human species do not share the same patterns of variation through geography. Skin color varies with latitude and certain people are tall or have brown hair. There is a statistical correlation between particular features in a population, but different features are not expressed or inherited together. Thus, genes which code for superficial physical traits—such as skin color, hair color, or height—represent a minuscule and insignificant portion of the human genome and do not correlate with genetic affinity. Dark-skinned populations that are found in Africa, Australia, and South Asia are not closely related to each other. Even within the same region, physical phenotype is not related to genetic affinity: dark-skinned Ethiopians are more closely related to light-skinned Armenians than to dark-skinned Bantu populations. Despite pygmy populations of South East Asia (Andamanese) having similar physical features with African pygmy populations such as short stature, dark skin, and curly hair, they are not genetically closely related to these populations. Genetic variants affecting superficial anatomical features (such as skin color)—from a genetic perspective, are essentially meaningless—they involve a few hundred of the billions of nucleotides in a person's DNA. Individuals with the same morphology do not necessarily cluster with each other by lineage, and a given lineage does not include only individuals with the same trait complex.

Due to practices of group endogamy, allele frequencies cluster locally around kin groups and lineages, or by national, ethnic, cultural and linguistic boundaries, giving a detailed degree of correlation between genetic clusters and population groups when considering many alleles simultaneously. Despite this, there are no genetic boundaries around local populations that biologically mark off any discrete groups of humans. Human variation is continuous, with no clear points of demarcation. There are no large clusters of relatively homogeneous people and almost every individual has genetic alleles from several ancestral groups.

Psychology

The human brain, the focal point of the central nervous system in humans, controls the peripheral nervous system. In addition to controlling "lower," involuntary, or primarily autonomic activities such as respiration and digestion, it is also the locus of "higher" order functioning such as thought, reasoning, and abstraction. These cognitive processes constitute the mind, and, along with their behavioral consequences, are studied in the field of psychology.

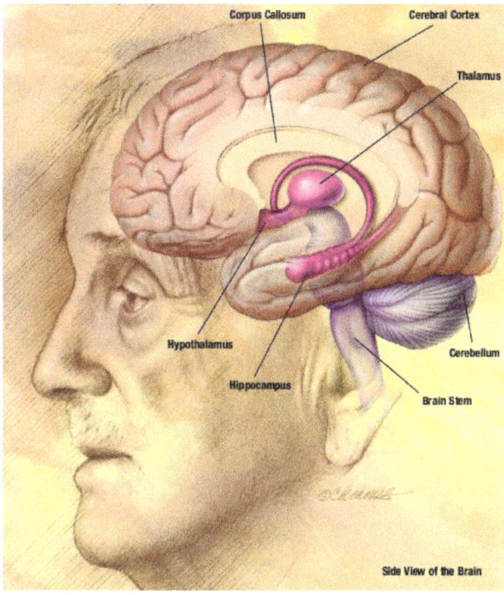

Drawing of the human brain, showing several important structures

Generally regarded as more capable of these higher order activities, the human brain is believed to be more "intelligent" in general than that of any other known species. While some non-human species are capable of creating structures and using simple tools—mostly through instinct and mimicry—human technology is vastly more complex, and is constantly evolving and improving through time.

Sleep and Dreaming

Humans are generally diurnal. The average sleep requirement is between seven and nine hours per day for an adult and nine to ten hours per day for a child; elderly people usually sleep for six to seven hours. Having less sleep than this is common among humans, even though sleep deprivation can have negative health effects. A sustained restriction of adult sleep to four hours per day has been shown to correlate with changes in physiology and mental state, including reduced memory, fatigue, aggression, and bodily discomfort. During sleep humans dream. In dreaming humans experience sensory images and sounds, in a sequence which the dreamer usually perceives more as an apparent participant than as an observer. Dreaming is stimulated by the pons and mostly occurs during the REM phase of sleep.

Consciousness and Thought

Humans are one of the relatively few species to have sufficient self-awareness to recognize themselves in a mirror. Already at 18 months, most human children are aware that the mirror image is not another person.

Humans in a lecture at the Faculty of Biomedical Engineering, CTU, in Prague.

The human brain perceives the external world through the senses, and each individual human is influenced greatly by his or her experiences, leading to subjective views of existence and the passage of time. Humans are variously said to possess consciousness, self-awareness, and a mind, which correspond roughly to the mental processes of thought. These are said to possess qualities such as self-awareness, sentience, sapience, and the ability to perceive the relationship between oneself and one's environment. The extent to which the mind constructs or experiences the outer world is a matter of debate, as are the definitions and validity of many of the terms used above.

The physical aspects of the mind and brain, and by extension of the nervous system, are studied in the field of neurology, the more behavioral in the field of psychology, and a sometimes loosely defined area between in the field of psychiatry, which treats mental illness and behavioral disorders. Psychology does not necessarily refer to the brain or nervous system, and can be framed purely in terms of phenomenological or information processing theories of the mind. Increasingly, however, an understanding of brain functions is being included in psychological theory and practice, particularly in areas such as artificial intelligence, neuropsychology, and cognitive neuroscience.

The nature of thought is central to psychology and related fields. Cognitive psychology studies cognition, the mental processes' underlying behavior. It uses information processing as a framework for understanding the mind. Perception, learning, problem solving, memory, attention, language and emotion are all well researched areas as well. Cognitive psychology is associated with a school of thought known as cognitivism, whose adherents argue for an information processing model of mental function, informed by positivism and experimental psychology. Techniques and models from cognitive psychology are widely applied and form the mainstay of psychological theories in many areas of both research and applied psychology. Largely focusing on the development of the human mind through the life span, developmental psychology seeks to understand how people come to perceive, understand, and act within the world and how these processes change as they age. This may focus on intellectual, cognitive, neural, social, or moral development.

Some philosophers divide consciousness into phenomenal consciousness, which is experience itself, and access consciousness, which is the processing of the things in experience. Phenomenal consciousness is the state of being conscious, such as when they say "I am conscious." Access consciousness is being conscious *of* something in relation to abstract concepts, such as when one says "I am conscious of these words." Various forms of access consciousness include awareness, self-awareness, conscience, stream of consciousness, Husserl's phenomenology, and intentionality. The concept of phenomenal consciousness, in modern history, according to some, is closely related to the concept of qualia. Social psychology links sociology with psychology in their shared study of the nature and causes of human social interaction, with an emphasis on how people think towards each other and how they relate to each other. The behavior and mental processes, both human and non-human, can be described through animal cognition, ethology, evolutionary psychology, and comparative psychology as well. Human ecology is an academic discipline that investigates how humans and human societies interact with both their natural environment and the human social environment.

Motivation and Emotion

Illustration of grief from Charles Darwin's book *The Expression of the Emotions in Man and Animals*.

Motivation is the driving force of desire behind all deliberate actions of humans. Motivation is based on emotion—specifically, on the search for satisfaction (positive emotional experiences), and the avoidance of conflict. Positive and negative is defined by the individual brain state, which may be influenced by social norms: a person may be driven to self-injury or violence because their brain is conditioned to create a positive response to these actions. Motivation is important because it is involved in the performance of all learned responses. Within psychology, conflict avoidance and the libido are seen to be primary motivators. Within economics, motivation is often seen to be based on incentives; these may be financial, moral, or coercive. Religions generally posit divine or demonic influences.

Happiness, or the state of being happy, is a human emotional condition. The definition of happiness is a common philosophical topic. Some people might define it as the best condition that a human can have—a condition of mental and physical health. Others define it as freedom from want and distress; consciousness of the good order of things; assurance of one's place in the universe or society.

Emotion has a significant influence on, or can even be said to control, human behavior, though historically many cultures and philosophers have for various reasons discouraged allowing this influence to go unchecked. Emotional experiences perceived as pleasant, such as love, admiration, or joy, contrast with those perceived as unpleasant, like hate, envy, or sorrow. There is often a distinction made between refined emotions that are socially learned and survival oriented emotions, which are thought to be innate. Human exploration of emotions as separate from other neurological phenomena is worthy of note, particularly in cultures where emotion is considered separate from physiological state. In some cultural medical theories emotion is considered so synonymous with certain forms of physical health that no difference is thought to exist. The Stoics believed excessive emotion was harmful, while some Sufi teachers felt certain extreme emotions could yield a conceptual perfection, what is often translated as ecstasy.

In modern scientific thought, certain refined emotions are considered a complex neural trait innate in a variety of domesticated and non-domesticated mammals. These were commonly developed in reaction to superior survival mechanisms and intelligent interaction with each other and the environment; as such, refined emotion is not in all cases as discrete and separate from natural neural function as was once assumed. However, when humans function in civilized tandem, it has been noted that uninhibited acting on extreme emotion can lead to social disorder and crime.

Sexuality and Love

Human parents continue caring for their offspring long after they are born.

For humans, sexuality has important social functions: it creates physical intimacy, bonds and hierarchies among individuals, besides ensuring biological reproduction. Sexual desire or libido, is experienced as a bodily urge, often accompanied by strong emotions such as love, ecstasy and jealousy. The significance of sexuality in the human species is reflected in a number of physical features among them hidden ovulation, the evolution of external scrotum and penis suggesting sperm competition, the absence of an os penis, permanent secondary sexual characteristics and the forming of pair bonds based on sexual attraction as a common social structure. Contrary to other primates that often advertise estrus through visible signs, human females do not have a distinct or visible signs of ovulation plus they experience sexual desire outside of their fertile periods. These adaptations indicate that the meaning of sexuality in humans is similar to that found in the bonobo, and that the complex human sexual behavior has a long evolutionary history.

Human choices in acting on sexuality are commonly influenced by cultural norms which vary widely. Restrictions are often determined by religious beliefs or social customs. The pioneering researcher Sigmund Freud believed that humans are born polymorphously perverse, which means that any number of objects could be a source of pleasure. According to Freud humans then pass through five stages of psychosexual development and can fixate on any stage because of various traumas during the process. For Alfred Kinsey, another influential sex researcher, people can fall anywhere along a continuous scale of sexual orientation, with only small minorities fully heterosexual or homosexual. Recent studies of neurology and genetics suggest people may be born predisposed to various sexual tendencies.

Behavior

Humans often live in family-based social structures.

Humans are highly social beings and tend to live in large complex social groups. More than any other creature, humans are capable of utilizing systems of communication for self-expression, the exchange of ideas, and organization, and as such have created complex social structures composed of many cooperating and competing groups. Human groups range from the size of families to nations. Social interactions between humans have established an extremely wide variety of values, social norms, and rituals, which together form the basis of human society.

Culture is defined here as patterns of complex symbolic behavior, i.e. all behavior that is not innate but which has to be learned through social interaction with others; such as the use of distinctive material and symbolic systems, including language, ritual, social organization, traditions, beliefs and technology.

Language

While many species communicate, language is unique to humans, a defining feature of humanity, and a cultural universal. Unlike the limited systems of other animals, human language is open – an infinite number of meanings can be produced by combining a limited number of symbols. Human language also has the capacity of displacement, using words to represent things and happenings that are not presently or locally occurring, but reside in the shared imagination of interlocutors. Language differs from other forms of communication in that it is modality independent; the same meanings can be conveyed through different media, auditively in speech, visually by sign language or writing, and even through tactile media such as braille. Language is central to the communication between humans, and to the sense of identity that unites nations, cultures and ethnic groups. The invention of writing systems at least five thousand years ago allowed the preservation of language on material objects, and was a major technological advancement. The science of linguistics describes the structure and function of language and the relationship between languages. There are approximately six thousand different languages currently in use, including sign languages, and many thousands more that are extinct.

Gender Roles

The sexual division of humans into male and female has been marked culturally by a corresponding division of roles, norms, practices, dress, behavior, rights, duties, privileges, status, and power. Cultural differences by gender have often been believed to have arisen naturally out of a division of reproductive labor; the biological fact that women give birth led to their further cultural responsibility for nurturing and caring for children. Gender roles have varied historically, and challenges to predominant gender norms have recurred in many societies.

Kinship

All human societies organize, recognize and classify types of social relationships based on relations between parents and children (consanguinity), and relations through marriage (affinity). These kinds of relations are generally called kinship relations. In most societies kinship places mutual responsibilities and expectations of solidarity on the individuals that are so related, and those who recognize each other as kinsmen come to form networks through which other social institutions can be regulated. Among the many functions of kinship is the ability to form descent groups, groups of people sharing a common line of descent, which can function as political units such as clans. Another function is the way in which kinship unites families through marriage, forming kinship alliances between groups of wife-takers and wife-givers. Such alliances also often have important political and economical ramifications, and may result in the formation of political organization above the community level. Kinship relations often includes regulations for whom an individual should or shouldn't marry. All societies have rules of incest taboo, according to which marriage between certain kinds of kin relations are prohibited – such rules vary widely between cultures. Some societies also have rules of preferential marriage with certain kin relations, frequently with either cross or parallel cousins. Rules and norms for marriage and social behavior among kinsfolk is often reflected in the systems of kinship terminology in the various languages of the world. In many societies kinship relations can also be formed through forms of co-habitation, adoption, fostering, or companionship, which also tends to create relations of enduring solidarity (nurture kinship).

Sessue Hayakawa (left) with actress and wife Tsuru Aoki in a screen shot of the 1919 film *The Dragon Painter*.

Ethnicity

Humans often form ethnic groups, such groups tend to be larger than kinship networks and be organized around a common identity defined variously in terms of shared ancestry and history, shared cultural norms and language, or shared biological phenotype. Such ideologies of shared characteristics are often perpetuated in the form of powerful, compelling narratives that give legitimacy and continuity to the set of shared values. Ethnic groupings often correspond to some level of political organization such as the band, tribe, city state or nation. Although ethnic groups appear and disappear through history, members of ethnic groups often conceptualize their groups as having histories going back into the deep past. Such ideologies give ethnicity a powerful role in defining social identity and in constructing solidarity between members of an ethno-political unit. This unifying property of ethnicity has been closely tied to the rise of the nation state as the predominant form of political organization in the 19th and 20th century.

Society, Government, and Politics

Society is the system of organizations and institutions arising from interaction between humans. A state is an organized political community occupying a definite territory, having an organized government, and possessing internal and external sovereignty. Recognition of the state's claim to independence by other states, enabling it to enter into international agreements, is often important to the establishment of its statehood. The "state" can also be defined in terms of domestic conditions, specifically, as conceptualized by Max Weber, "a state is a human community that (successfully) claims the monopoly of the 'legitimate' use of physical force within a given territory."

The United Nations Headquarters in New York City, which houses one of the largest political organizations in the world

Russian honor guard at Tomb of the Unknown Soldier.

Government can be defined as the political means of creating and enforcing laws; typically via a bureaucratic hierarchy. Politics is the process by which decisions are made within groups; this process often involves conflict as well as compromise. Although the term is generally applied to behavior within governments, politics is also observed in all human group interactions, including corporate, academic, and religious institutions. Many different political systems exist, as do many different ways of understanding them, and many definitions overlap. Examples of governments include monarchy, Communist state, military dictatorship, theocracy, and liberal democracy, the last of which is considered dominant today. All of these issues have a direct relationship with economics.

Trade and Economics

Trade is the voluntary exchange of goods and services, and is a form of economics. A mechanism that allows trade is called a market. Modern traders instead generally negotiate through a medium of exchange, such as money. As a result, buying can be separated from selling, or earning. Because of specialization and division of labor, most people concentrate on a small aspect of manufacturing or service, trading their labor for products. Trade exists between regions because different regions have an absolute or comparative advantage in the production of some tradable commodity, or because different regions' size allows for the benefits of mass production.

Buyers and sellers bargaining in a market

Economics is a social science which studies the production, distribution, trade, and consumption of goods and services. Economics focuses on measurable variables, and is broadly divided into two main branches: microeconomics, which deals with individual agents, such as households and businesses, and macroeconomics, which considers the economy as a whole, in which case it considers aggregate supply and demand for money, capital and commodities. Aspects receiving particular attention in economics are resource allocation, production, distribution, trade, and competition. Economic logic is increasingly applied to any problem that involves choice under scarcity or determining economic value.

War is a state of organized armed conflict between states or non-state actors. War is characterized by the use of lethal violence against others – whether between combatants or upon non-combatants – to achieve military goals through force. Lesser, often spontaneous conflicts, such as brawls, riots, revolts, and melees, are not considered to be warfare. Revolutions can be nonviolent or an organized and armed revolution which denotes a state of war. During the 20th century, it is estimated that between 167 and 188 million people died as a result of war. A common definition defines war as a series of military campaigns between at least two opposing sides involving a dispute over sovereignty, territory, resources, religion, or other issues. A war between internal elements of a state is a civil war. Among animals, all-out war against fellow members of the same species occurs only among large societies of humans and ants.

The mushroom cloud of the atomic bombing of Nagasaki on August 9, 1945.

There have been a wide variety of rapidly advancing tactics throughout the history of war, ranging from conventional war to asymmetric warfare to total war and unconventional warfare. Techniques include hand to hand combat, the use of ranged weapons, naval warfare, and, more recently, air support. Military intelligence has often played a key role in determining victory and defeat. Propaganda, which often includes information, slanted opinion and disinformation, plays a key role both in maintaining unity within a warring group and in sowing discord among opponents. In modern warfare, soldiers and combat vehicles are used to control the land, warships the sea, and aircraft the sky. These fields have also overlapped in the forms of marines, paratroopers, aircraft carriers, and surface-to-air missiles, among others. Satellites in low Earth orbit have made outer space a factor in warfare as well as it is used for detailed intelligence gathering, however no known aggressive actions have been taken from space.

Soldiers in front of the wood of Hougoumont during the reenactment of the battle of Waterloo (1815), June 2011, Waterloo, Belgium.

Material Culture and Technology

Stone tools were used by proto-humans at least 2.5 million years ago. The controlled use of fire began around 1.5 million years ago. Since then, humans have made major advances, developing complex technology to create tools to aid their lives and allowing for other advancements in culture. Major leaps in technology include the discovery of agriculture – what is known as the Neolithic Revolution, and the invention of automated machines in the Industrial Revolution.

An array of Neolithic artifacts, including bracelets, axe heads, chisels, and polishing tools.

Archaeology attempts to tell the story of past or lost cultures in part by close examination of the artifacts they produced. Early humans left stone tools, pottery, and jewelry that are particular to various regions and times.

Body Culture

Throughout history, humans have altered their appearance by wearing clothing and adornments, by trimming or shaving hair or by means of body modifications.

Body modification is the deliberate altering of the human body for any non-medical reason, such as aesthetics, sexual enhancement, a rite of passage, religious reasons, to display group membership or affiliation, to create body art, shock value, or self-expression. In its most broad definition it includes plastic surgery, socially acceptable decoration (e.g. common ear piercing in many societies), and religious rites of passage (e.g. circumcision in a number of cultures).

Philosophy and Self-Reflection

Philosophy is a discipline or field of study involving the investigation, analysis, and development of ideas at a general, abstract, or fundamental level. It is the discipline searching for a general understanding of reality, reasoning and values. Major fields of philosophy include logic, metaphysics, epistemology, philosophy of mind, and axiology (which includes ethics and aesthetics). Philosophy covers a very wide range of approaches, and is used to refer to a worldview, to a perspective on an issue, or to the positions argued for by a particular philosopher or school of philosophy.

Statue of Confucius on Chongming Island in Shanghai

Religion and Spirituality

Religion is generally defined as a belief system concerning the supernatural, sacred or divine, and practices, values, institutions and rituals associated with such belief. Some religions also have a moral code. The evolution and the history of the first religions have recently become areas of active scientific investigation. However, in the course of its development, religion has taken on many forms that vary by culture and individual perspective. Some of the chief questions and issues religions are concerned with include life after death (commonly involving belief in an afterlife), the origin of life, the nature of the universe (religious cosmology) and its ultimate fate (eschatology), and what is moral or immoral. A common source for answers to these questions are beliefs in transcendent divine beings such as deities or a singular God, although not all religions are theistic. Spirituality, belief or involvement in matters of the soul or spirit, is one of the many different approaches humans take in trying to answer fundamental questions about humankind's place in the universe, the meaning of life, and the ideal way to live one's life. Though these topics have also been addressed by philosophy, and to some extent by science, spirituality is unique in that it focuses on mystical or supernatural concepts such as karma and God.

Although the exact level of religiosity can be hard to measure, a majority of humans professes some variety of religious or spiritual belief, although many (in some countries a majority) are irreligious. This includes humans who have no religious beliefs or do not identify with any religion. Humanism is a philosophy which seeks to include all of humanity and all issues common to humans; it is usually non-religious. Most religions and spiritual beliefs are clearly distinct from science on both a philosophical and methodological level; the two are not generally considered mutually exclusive and a majority of humans hold a mix of both scientific and religious views. The distinction between philosophy and religion, on the other hand, is at times less clear, and the two are linked in such fields as the philosophy of religion and theology.

Art, Music, and Literature

Art is a cultural universal, and humans have been producing artistic works at least since the days of Cro Magnon. Art may be defined as a form of cultural expression and the usage of narratives of liberation and exploration (i.e. art history, art criticism, and art theory) to mediate its boundaries. This distinction may be applied to objects or performances, current or historical, and its prestige extends to those who made, found, exhibit, or own them. In the modern use of the word, art is commonly understood to be the process or result of making material works that, from concept to creation, adhere to the "creative impulse" of human beings. Art is distinguished from other works by being in large part unprompted by necessity, by biological drive, or by any undisciplined pursuit of recreation.

Allegory of Music (ca. 1594), a painting of a woman writing sheet music by Lorenzo Lippi

Music is a natural intuitive phenomenon based on the three distinct and interrelated organization structures of rhythm, harmony, and melody. Listening to music is perhaps the most common and universal form of entertainment, while learning and understanding it are popular disciplines. There are a wide variety of music genres and ethnic musics. Literature, the body of written—and possibly oral—works, especially creative ones, includes prose, poetry and drama, both fiction and non-fiction. Literature includes such genres as epic, legend, myth, ballad, and folklore.

Science

Another unique aspect of human culture and thought is the development of complex methods for acquiring knowledge through observation, quantification, and verification. The scientific method has been developed to acquire knowledge of the physical world and the rules, processes and principles of which it consists, and combined with mathematics it enables the prediction of complex patterns of causality and consequence. Some other animals are able to recognize differences in small quantities, but humans are able to understand and recognize much larger, even abstract, quantities, and to recognize and understand algorithmic patterns which enables infinite counting routines and algebra, something that is not found in any other species.

Soil Health

Soil health is a state of a soil meeting its range of ecosystem functions as appropriate to its environment. Soil health testing is an assessment of this status. Soil health depends on soil biodiversity (with a robust soil biota), and it can be improved via soil conditioning (soil amendment).

Aspects

The term soil health is used to describe the state of a soil in:

- Sustaining plant and animal productivity and biodiversity (Soil biodiversity);
- Maintaining or enhance water and air quality;
- Supporting human health and habitation.

Soil Health has partly if not largely replaced the expression "Soil Quality" that was extant in the 1990s. The primary difference between the two expressions is that soil quality was focused on individual traits within a functional group, as in "quality of soil for maize production" or "quality of soil for roadbed preparation" and so on. The addition of the word "health" shifted the perception to be integrative, holistic and systematic. The two expressions still overlap considerably.

The underlying principle in the use of the term "soil health" is that soil is not just an inert, lifeless growing medium, which modern farming tends to represent, rather it is a living, dynamic and ever-so-subtly changing whole environment. It turns out that soils highly fertile from the point of view of crop productivity are also lively from a biological point of view. It is now commonly recognized that soil microbial biomass is large: in temperate grassland soil the bacterial and fungal biomass have been documented to be 1–2 t (2.0 long tons; 2.2 short tons)/hectare and 2–5 t (4.9 long tons; 5.5 short tons)/ha, respectively. Some microbiologists now believe that 80% of soil nutrient functions are essentially controlled by microbes. If this is consistently true, than the prevailing Liebig nutrient theory model, which excludes biology, is perhaps dangerously incorrect for managing soil fertility sustainably for the future.

Using the human health analogy, a healthy soil can be categorized as one:

- In a state of composite well-being in terms of biological, chemical and physical properties;
- Not diseased or infirmed (i.e. not degraded, nor degrading), nor causing nega-

tive off-site impacts;

- With each of its qualities cooperatively functioning such that the soil reaches its full potential and resists degradation;
- Providing a full range of functions (especially nutrient, carbon and water cycling) and in such a way that it maintains this capacity into the future.

Conceptualisation

Soil health is the condition of the soil in a defined space and at a defined scale relative to a set of benchmarks that encompass healthy functioning. It would not be appropriate to refer to soil health for soil-roadbed preparation, as in the analogy of soil quality in a functional class. The definition of soil health may vary between users of the term as alternative users may place differing priorities upon the multiple functions of a soil. Therefore, the term soil health can only be understood within the context of the user of the term, and their aspirations of a soil, as well as by the boundary definition of the soil at issue.

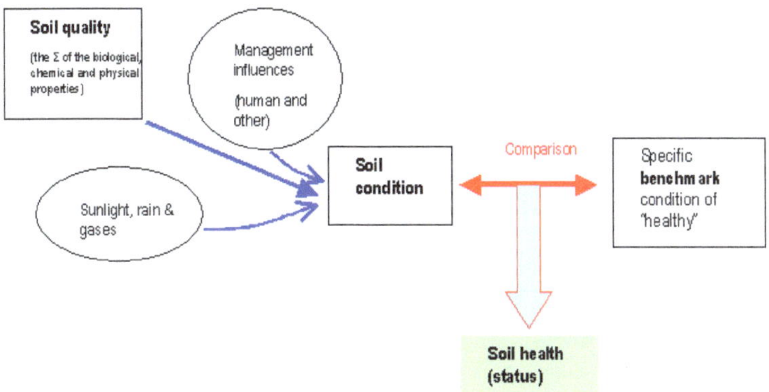

Interpretation

Different soils will have different benchmarks of health depending on the "inherited" qualities, and on the geographic circumstance of the soil. The generic aspects defining a healthy soil can be considered as follows:

- "Productive" options are broad;
- Life diversity is broad;
- Absorbency, storing, recycling and processing is high in relation to limits set by climate;
- Water runoff quality is of high standard;
- Low entropy; and,
- No damage to, or loss of the fundamental components.

This translates to:

- A comprehensive cover of vegetation;
- Carbon levels relatively close to the limits set by soil type and climate;
- Little leakage of nutrients from the ecosystem;
- Biological and agricultural productivity relatively close to the limits set by the soil environment and climate;
- Only geological rates of erosion;
- No accumulation of contaminants; and,
- The ecosystem does not rely excessively on inputs of fossil energy

An unhealthy soil thus is the simple converse of the above.

Measurement

On the basis of the above, soil health will be measured in terms of individual ecosystem services provided relative to the benchmark. Specific benchmarks used to evaluate soil health include CO_2 release, humus levels, microbial activity, and available calcium.

Soil health testing is spreading in the USA, Australia and S. Africa. Cornell University has had a Soil Health Test since 2006 and offers many types of individual soil health tests as well as three different soil health testing packages. The approach is to add into common chemical nutrient testing a biological set of factors not normally included in routine soil testing. The best example is adding biological soil respiration (CO2-Burst) as a test procedure; this has already been adapted to modern commercial labs in the past 10 years.

There is resistance among soil testing labs and university scientists to adding new biological tests, primarily since interpretation of soil fertility is based on models from "crop response" studies which match yield to test levels of specific chemical nutrients. These soil test methods have evolved slowly over the past 40 years and are backed by considerable effort. However, in this same time USA soils have also lost up to 75% of their carbon (humus), causing biological fertility to drop drastically. Many critics of the current system say this is sufficient evidence that the old soil testing models have failed us, and need to be replaced with new approaches. These older models have stressed "maximum yield" and " yield calibration" to such an extent that related factors have been overlooked. Thus, surface and groundwater pollution with excess nutrients (nitrates and phosphates) has grown enormously, and is reported presently (in the USA) to be the worst it has been since the 1970s, before the advent of environmental consciousness.

Soil pH

The soil pH is a measure of the acidity or alkalinity in soils. pH is defined as the negative logarithm (base 10) of the activity of hydronium ions (H+ or, more precisely, H 3O+ aq) in a solution. In water, it normally ranges from -1 to 14, with 7 being neutral. A pH below 7 is acidic and above 7 is alkaline. Soil pH is considered a master variable in soils as it controls many chemical processes that take place. It specifically affects plant nutrient availability by controlling the chemical forms of the nutrient. The optimum pH range for most plants is between 5.5 and 7.0, however many plants have adapted to thrive at pH values outside this range.

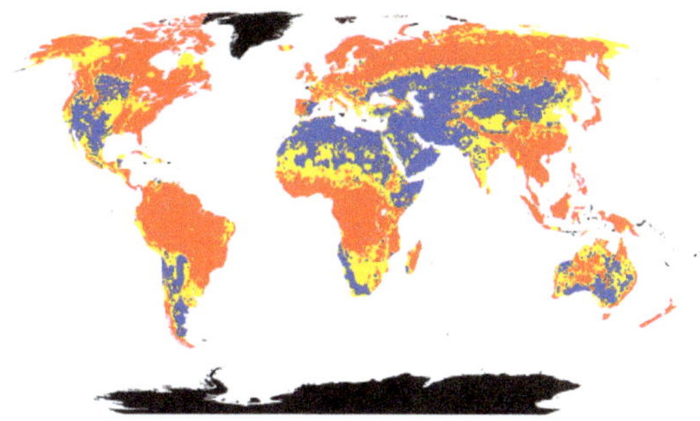

Global variation in soil pH. Red = acidic soil. Yellow = neutral soil. Blue = alkaline soil. Black = no data.

Classification of Soil pH Ranges

The United States Department of Agriculture Natural Resources Conservation Service, formerly Soil Conservation Service classifies soil pH ranges as follows:

Denomination	pH range
Ultra acid	< 3.5
Extremely acid	3.5–4.4
Very strongly acid	4.5–5.0
Strongly acid	5.1–5.5
Moderately acid	5.6–6.0
Slightly acid	6.1–6.5
Neutral	6.6–7.3
Slightly alkaline	7.4–7.8
Moderately alkaline	7.9–8.4
Strongly alkaline	8.5–9.0
Very strongly alkaline	> 9.0

Sources of Soil pH

Sources of Acidity

Acidity in soils comes from H^+ and Al^{3+} ions in the soil solution and sorbed to soil surfaces. While pH is the measure of H^+ in solution, Al^{3+} is important in acid soils because between pH 4 and 6, Al^{3+} reacts with water (H_2O) forming $AlOH^{2+}$, and $Al(OH)_2^+$, releasing extra H^+ ions. Every Al^{3+} ion can create 3 H^+ ions. Many other processes contribute to the formation of acid soils including rainfall, fertilizer use, plant root activity and the weathering of primary and secondary soil minerals. Acid soils can also be caused by pollutants such as acid rain and mine spoilings.

- Rainfall: Acid soils are most often found in areas of high rainfall. Excess rainfall leaches base cation from the soil, increasing the percentage of Al^{3+} and H^+ relative to other cations. Additionally, rainwater has a slightly acidic pH of 5.7 due to a reaction with CO_2 in the atmosphere that forms carbonic acid.
- Fertilizer use: Ammonium (NH_4^+) fertilizers react in the soil in a process called nitrification to form nitrate (NO_3^-), and in the process release H^+ ions.
- Plant root activity: Plants take up nutrients in the form of ions (NO_3^-, NH_4^+, Ca^{2+}, $H_2PO_4^-$, etc.), and often, they take up more cations than anions. However plants must maintain a neutral charge in their roots. In order to compensate for the extra positive charge, they will release H^+ ions from the root. Some plants will also exude organic acids into the soil to acidify the zone around their roots to help solubilize metal nutrients that are insoluble at neutral pH, such as iron (Fe).
- Weathering of minerals: Both primary and secondary minerals that compose soil contain Al. As these minerals weather, some components such as Mg, Ca, and K, are taken up by plants, others such as Si are leached from the soil, but due to chemical properties, Fe and Al remain in the soil profile. Highly weathered soils are often characterized by having high concentrations of Fe and Al oxides.
- Acid rain: When atmospheric water reacts with sulfur and nitrogen compounds that result from industrial processes, the result can be the formation of sulfuric and nitric acid in rainwater. However the amount of acidity that is deposited in rainwater is much less, on average, than that created through agricultural activities.
- Mine spoil: Severely acidic conditions can form in soils near mine spoils due to the oxidation of pyrite.
- Potential acid sulfate soils naturally formed in waterlogged coastal and estuarine environments can become highly acidic when drained or excavated.

- Decomposition of organic matter by microorganisms releases CO_2 which when mixed with soil water can form carbonic acid (H_2CO_3).

Sources of Alkalinity

Alkaline soils have a high saturation of base cations (K^+, Ca^{2+}, Mg^{2+} and Na^+). This is due to an accumulation of soluble salts which are classified as either saline soil, sodic soil, saline-sodic soil or alkaline soil. All saline and sodic soils have high salt concentrations, with saline soils being dominated by calcium and magnesium salts and sodic soils being dominated by sodium. Alkaline soils are characterized by the presence of carbonates. Soil in areas with limestone near the surface are alkaline from the calcium carbonate in limestone constantly mixing with the soil. Groundwater sources in these areas contain dissolved limestone.

Effect of Soil pH on Plant Growth

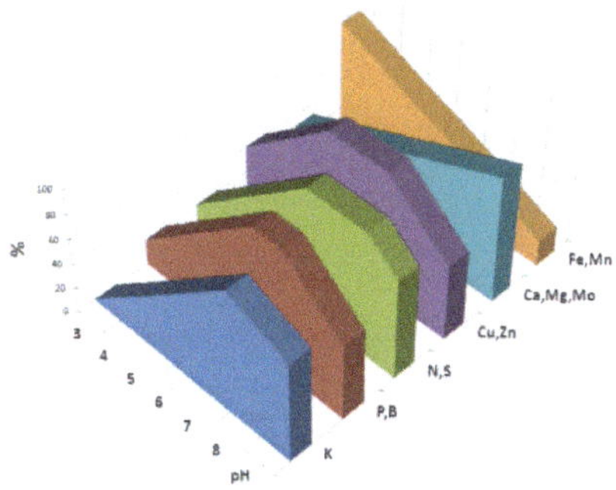

Nutrient availability in relation to soil pH

Acid Affected Soils

Plants grown in acid soils can experience a variety of symptoms including aluminium (Al), hydrogen (H), and/or manganese (Mn) toxicity, as well as nutrient deficiencies of calcium (Ca) and magnesium (Mg).

Aluminium toxicity is the most widespread problem in acid soils. Aluminium is present in all soils, but dissolved Al^{3+} is toxic to plants; Al^{3+} is most soluble at low pH, above pH 5.2 little Al is in soluble form in most soils. Aluminium is not a plant nutrient, and as such, is not actively taken up by the plants, but enters plant roots passively through osmosis. Aluminium inhibits root growth; lateral roots and root tips become thickened and roots lack fine branching; root tips may turn brown. In the root, Al has been shown to interfere with many physiological processes including the uptake and transport of calcium and other essential nutrients, cell division, cell wall formation, and enzyme activity.

Below pH 4, H^+ ions themselves damage root cell membranes.

In soils with high content of manganese-containing minerals, Mn toxicity can become a problem at pH 5.6 and lower. Manganese, like aluminium, becomes increasingly soluble as pH drops, and Mn toxicity symptoms can be seen at pH levels below 5.6. Manganese is an essential plant nutrient, so plants transport Mn into leaves. Classic symptoms of Mn toxicity are crinkling or cupping of leaves.

Nutrient Availability in Relation to Soil pH

Nutrients needed in large amounts by plants are referred to as macronutrients and include nitrogen (N), phosphorus (P), potassium (K), calcium (Ca), magnesium (Mg) and sulfur (S). Elements that plants need in trace amounts are called trace nutrients or micronutrients. Trace nutrients are not major components of plant tissue but are essential for growth. They include iron (Fe), manganese (Mn), zinc (Zn), copper (Cu), cobalt (Co), molybdenum (Mo), and boron (B). Both macronutrient and micronutrient availability are affected by soil pH. In slightly to moderately alkaline soils, molybdenum and macronutrient (except for phosphorus) availability is increased, but P, Fe, Mn, Zn Cu, and Co levels are reduced and may adversely affect plant growth. In acidic soils, micronutrient availability (except for Mo and Bo) is increased. Nitrogen is supplied as ammonium (NH4) or nitrate (NO3) by nitrogen fixation or fertilizer amendments, and dissolved N will have the highest concentrations in soil with pH 6–8. Concentrations of available N are less sensitive to pH than concentration of available P. In order for P to be available for plants, soil pH needs to be in the range 6.0 and 7.5. If pH is lower than 6, P starts forming insoluble compounds with iron (Fe) and aluminium (Al) and if pH is higher than 7.5 P starts forming insoluble compounds with calcium (Ca). Most nutrient deficiencies can be avoided between a pH range of 5.5 to 6.5, provided that soil minerals and organic matter contain the essential nutrients to begin with.

Water Availability in Relation to Soil pH

Determining pH

Methods of determining pH include:

- Observation of soil profile: Certain profile characteristics can be indicators of either acid, saline, or sodic conditions. Strongly acidic soils often have poor incorporation of the organic surface layer with the underlying mineral layer. The mineral horizons are distinctively layered in many cases, with a pale eluvial (E) horizon beneath the organic surface; this E is underlain by a darker B horizon in a classic podzol horizon sequence. This is a very rough gauge of acidity as there is no correlation between thickness of the E and soil pH. E horizons a few feet thick in Florida usually have pH just above 5 (merely "strongly acid") while E horizons a few inches thick in New England are "extremely acid" with pH readings of 4.5 or below. In the southern Blue Ridge Mountains there are "ultra acid" soils, pH below 3.5, which have no E horizon. Presence of a caliche layer indicates the presence of calcium carbonates, which are present in alkaline conditions. Also, columnar structure can be an indicator of sodic condition.

- Observation of predominant flora. Calcifuge plants (those that prefer an acidic soil) include *Erica*, *Rhododendron* and nearly all other Ericaceae species, many birch (*Betula*), foxglove (*Digitalis*), gorse (*Ulex* spp.), and Scots Pine (*Pinus sylvestris*). Calcicole (lime loving) plants include ash trees (*Fraxinus* spp.), honeysuckle (*Lonicera*), *Buddleja*, dogwoods (*Cornus* spp.), lilac (*Syringa*) and *Clematis* species.
- Use of an inexpensive pH testing kit, where in a small sample of soil is mixed with indicator solution which changes colour according to the acidity/alkalinity.
- Use of litmus paper. A small sample of soil is mixed with distilled water, into which a strip of litmus paper is inserted. If the soil is acidic the paper turns red, if alkaline, blue.
- Use of a commercially available electronic pH meter, in which a rod is inserted into moistened soil and measures the concentration of hydrogen ions.

Examples of Plant pH Preferences

- pH 4.5–5.0: Ericaceae (azalea, bilberry, blueberry, cranberry; heather), hydrangea for blue, (less acidic for pink), liquidambar or sweet gum, orchid, pin oak
- pH 5.0–5.5: Boronia, daphne; Ericaceae: (camellia, heather, rhododendron), ferns, iris, orchids, parsley, conifers (e.g., pine); Poaceae: (maize, millet, rye, oat), radish; Solanales: (potato, sweet potato); Bromeliaceae (pineapple).
- pH 5.5–6.0: Asteraceae: (aster, endive); Brassicaceae: (brussels sprout, kohlrabi), carrot; Cucurbitales: (begonia, chayote or choko); Fabaceae: (bean, crimson clover, peanut, soybean), petunia, rhubarb, violet, most bulbs (canna, daffodil, jonquil), larkspur, primrose
- pH 6.0–6.5 antirrhinum or snapdragon, Brassicaceae: (broccoli, cabbage, candytuft, cauliflower, turnip, wallflower); Cucurbitaceae: (cucumber, pumpkin, squash); Fabaceae: (pea, red clover, white clover), gladiolus, Iceland poppy; Rosales: (cannabis, rose, strawberry); Solanaceae: (eggplant or aubergine, tomato), sweet corn; Violaceae: (pansy, viola), zinnia or zinnea
- pH 6.5–7.0: Amaranthaceae: (beet, spinach); Apiaceae: (celery, parsnip); Asparagales: (asparagus, onion); Asteraceae: (chrysanthemum, dahlia, lettuce), carnation; Fabaceae: (alfalfa, sweet pea), melon, stock, tulip
- pH 7.1–8.0 lilac

Changing Soil pH

Increasing pH of Acidic Soil

The most common amendment to increase soil pH is lime ($CaCO_3$ or $MgCO_3$), usually in the form of finely ground agricultural lime. The amount of lime needed to change pH is determined by the mesh size of the lime (how finely it is ground) and the buffering capacity of the soil. A high mesh size (60–100) indicates a finely ground lime, that will react quickly with soil acidity. Buffering capacity of soils is a function of a soils cation exchange capacity, which is in turn determined by the clay content of the soil, the type of clay and the amount of organic matter present. Soils with high clay content, particularly shrink–swell clay, will have a higher buffering capacity than soils with little clay. Soils with high organic matter will also have a higher buffering capacity than those with low organic matter. Soils with high buffering capacity require a greater amount of lime to be added than a soil with a lower buffering capacity for the same incremental change in pH. Other amendments that can be used to increase the pH of soil include wood ash, industrial CaO (burnt lime), and oyster shells. White firewood ash includes metal salts which are important for processes requiring ions such as Na^+ (sodium), K^+ (potassium), Ca^{2+} (calcium), which may or may not be good for the select flora, but decreases the acidic quality of soil. These products increase the pH of soils through the reaction of CO_3^{2-} with H^+ to produce CO_2 and H_2O. Calcium silicate neutralizes active acidity in the soil by removing free hydrogen ions, thereby increasing pH. As its silicate anion captures H^+ ions (raising the pH), it forms monosilicic acid (H_4SiO_4), a neutral solute.

Decreasing pH of Alkaline Soil

- Iron sulfates or aluminium sulfate as well as elemental sulfur (S) reduce pH through the formation of sulfuric acid.
- Urea, urea phosphate, ammonium nitrate, ammonium phosphates, ammonium sulfate and monopotassium phosphate fertilizers.
- organic matter in the form of plant litter, compost, and manure will decrease soil pH through the decomposition process. Certain acid organic matter such as pine needles, pine sawdust and acid peat are effective at reducing pH.

Soil Organic Matter

Soil organic matter (SOM) is the organic matter component of soil, consisting of plant and animal residues at various stages of decomposition, cells and tissues of soil organisms, and substances synthesized by soil organisms. SOM exerts numerous positive effects on soil physical and chemical properties, as well as the soil's capacity to provide regulatory ecosystem services. Particularly, the presence of SOM is regarded as being critical for soil function and soil quality.

The positive impacts of SOM result from a number of complex, interactive edaphic factors; a non-exhaustive list of SOM's effects on soil functioning includes improvements related to soil structure, aggregation, water retention, soil biodiversity, absorption and retention of pollutants, buffering capacity, and the cycling and storage of plant nutrients. SOM increases soil fertility by providing cation exchange sites and acting as reserve of plant nutrients, especially nitrogen (N), phosphorus (P), and sulfur (S), along with micronutrients, which are slowly released upon SOM mineralization. As such, there is a significant correlation between SOM content and soil fertility.

SOM also acts as a major sink and source of soil carbon (C). Although the C content of SOM is known to vary considerably, SOM is typically estimated to contain 58% C, and the terms 'soil organic carbon' (SOC) and SOM are often used interchangeably, with measured SOC content often serving as a proxy for SOM. Soil represents one of the largest C sinks on the planet and plays a major role in the global carbon cycle. Therefore, SOM/SOC dynamics and the capacity of soils to provide the ecosystem service of carbon sequestration through SOM management have received considerable attention in recent years.

The concentration of SOM in soils generally ranges from 1% to 6% of the total topsoil mass for most upland soils. Soils whose upper horizons consist of less than 1% organic matter are mostly limited to desert areas, while the SOM content of soils in low-lying, wet areas can be as high as 90%. Soils containing 12-18% SOC are generally classified as organic soils.

It can be divided into three general pools: living biomass of microorganisms, fresh and partially decomposed residues, and humus: the well-decomposed organic material. Surface plant litter is generally not included as part of soil organic matter.

Sources of Soil Organic Matter

The primary source of organic matter contained in soil is vegetal. In forest or prairies, as well as agricultural fields, dead plants are transformed by different kinds of living organisms. This process involves several steps, the first being mostly mechanical, and becoming more chemical as it progresses. The small living beings that work on that decomposition chain are themselves part of the soil organic matter, and form a food web of organisms that prey upon each other and are preyed upon.

There are also other animals that consume living vegetal material, whose residues are passed to the soil. The products from the living organisms metabolism are the secondary sources of soil organic matter that also includes the dead corpses of these organisms. Some animals, like earthworms, ants and centipedes contribute to both vertical and horizontal translocation of organic material.

Additional sources of soil organic matter include plant root exudates and charcoal.

Composition of Plant Residues

The water content of most plant residues is in the range of 60% to 90%. The dry matter consists of complex organic material composed mainly of carbon, oxygen and hydrogen. Although these three elements make up about 92% of the dry weight of the organic material in soils, there are other elements that are of great importance for the nutrition of plants. They include nitrogen, sulfur, phosphorus, potassium, calcium, magnesium and a range of micronutrients.

Organic compounds found in plant residues include:

- Carbohydrates are made up of carbon, hydrogen and oxygen, and range in complexity from rather simple *sugars* to the big molecules of *cellulose*.
- Fats consist of glycerids of fatty acids, like butyric, stearic, oleic. They are also made up of carbon, oxygen and hydrogen atoms.
- Lignins are complex compounds that form the older parts of wood, and consist also mainly of carbon, oxygen and hydrogen. They are resistant to decomposition.
- Proteins contain nitrogen in addition to carbon, hydrogen and oxygen, and also small amounts of sulfur, iron, and phosphorus.
- Charcoal is elemental carbon derived from incomplete combustion of organic matter. Charcoal is resistant to decomposition.

Decomposition

The vegetal residues in general are not water-soluble, and they are not usable by the plants. They constitute, nevertheless, the raw materials from which plant nutrients are derived. The decomposition is carried out as enzymatic biochemical processes by soil microorganisms, that obtain the necessary energy from the same residues, and produce the mineral compounds that are apt to be absorbed by plant roots. This process by which organic compounds are broken down and transformed into mineral (inorganic) compounds is also referred to as *mineralization*. A portion of organic material is not mineralized, but transformed into stable organic matter *humus*.

The break down of the organic compounds is done at very different rates, depending on their nature. The ranking, from fast to slow rates are as follow.

1. Sugars, starches and simple proteins.
2. Proteins

3. Hemicelluloses
4. Cellulose
5. Lignins and fats.

The reactions that take place can be included in one of three groups:

- Enzymatic oxidation, that produces carbon dioxide, water and heat. It affects the bulk or major portion of the material.
- The essential elements, nitrogen, sulfur, phosphorus are liberated and mineralized by a series of specific reactions.
- Compounds that are resistant to microbial action are formed by modification of original compounds or by synthesis of new ones by microbes, creating humus.

The list of mineral end products are as follow:

Element	Mineral end products
Carbon	CO_2, CO_3^{2-}, HCO_3^-, CH_4, C
Nitrogen	NH_4^+, NO_2^-, NO_3^-, N_2 (gas)
Sulfur	S, H_2S, SO_3^{2-}, SO_4^{2-}, CS_2
Phosphorus	$H_2PO_4^-$, HPO_4^{2-}
Others	H_2O, O_2, H_2, H^+, OH^-, K^+, Ca^{2+}, Mg^{2+}, etc.

Humus

As vegetal material undergoes decomposition, some microbial resistant compounds are formed. These include modified lignins, oils, fats and waxes. Secondly, some new compounds are synthesized, like polysaccharides and polyuronids. These materials form the basis for humus. New reactions take place between these compounds and some proteins and other nitrogen containing products, incorporating thus nitrogen and avoiding its mineralization. Other nutrients are also protected in this way from mineralization.

Humic substances classification. There is a classification into three groups, based on solubility in acids and alkalis, and also related to stability.

-*Fulvic acid* is the group which contains the materials that have the lowest molecular weight, and are soluble in acids and alkali, and susceptible to microbial attack.

-*Humic acid* group contains the intermediate materials, with medium molecular weight, soluble in alkali, but insoluble in acid, and intermediate resistance to microbial attack.

-*Humin* is the generic name for the materials with highest molecular weight, that are darkest in color, insoluble in acid and alkali, and with the most resistance to microbial attack.

Role in Carbon Cycling

Soil plays a major role in the global carbon cycle, with the global soil carbon pool estimated at 2500 gigatons. This is 3.3 times the size of the atmospheric pool (750 gigatons) and 4.5 times the biotic pool (560 gigatons). The pool of organic carbon, which occurs primarily in the form of SOM, accounts roughly 1550 gigatons of the total global C pool, with the remainder accounted for by soil inorganic carbon (SIC). The pool of organic C exists in dynamic equilibrium between gains and losses; soil may therefore serve as either a sink or source of C, through sequestration or greenhouse gas emissions, respectively, depending on exogenous factors.

Nitrogen Fixation

Nitrogen fixation is a process in which nitrogen (N_2) in the atmosphere is converted into ammonia (NH_3). Atmospheric nitrogen or molecular dinitrogen (N_2) is relatively inert: it does not easily react with other chemicals to form new compounds. The fixation process frees nitrogen atoms from their triply bonded diatomic form, N≡N, to be used in other ways.

Nitrogen fixation, natural and synthetic, is essential for all forms of life because nitrogen is required to biosynthesize basic building blocks of plants, animals and other life forms, e.g., nucleotides for DNA and RNA, the coenzyme nicotinamide adenine dinucleotide for its role in metabolism (transferring electrons between molecules), and amino acids for proteins. Therefore, as part of the nitrogen cycle, it is essential for agriculture and the manufacture of fertilizer. It is also an important process in the manufacture of explosives (e.g. gunpowder, dynamite, TNT, etc.). Nitrogen fixation occurs naturally in the soil by nitrogen fixing bacteria affiliated with some plants (for example, *Azotobacter* and legumes). Some nitrogen-fixing bacteria have very close relationships with plants, referred to as symbiotic nitrogen fixation. Looser relationships between nitrogen-fixing bacteria and plants are often referred to as associative or non-symbiotic, as seen in nitrogen fixation occurring on rice roots. It also occurs naturally in the air by means of lightning.

All biological nitrogen fixation is done by way of nitrogenase metalloenzymes which contain iron, molybdenum, or vanadium. Microorganisms that can fix nitrogen are prokaryotes (both bacteria and archaea, distributed throughout their respective kingdoms) called diazotrophs. Some higher plants, and some animals (termites), have formed associations (symbiosis) with diazotrophs.

Biological Nitrogen Fixation

Biological nitrogen fixation was discovered by the German agronomist Hermann Hellriegel and Dutch microbiologist Martinus Beijerinck. Biological nitrogen fixation (BNF) occurs when atmospheric nitrogen is converted to ammonia by an enzyme called a nitrogenase. The overall reaction for BNF is:

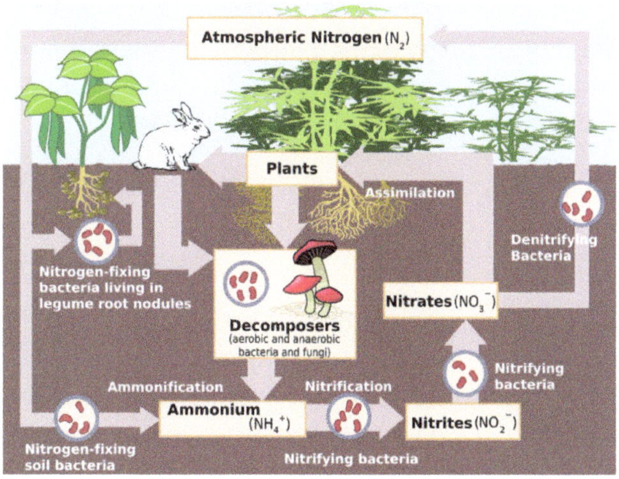

Schematic representation of the nitrogen cycle. Abiotic nitrogen fixation has been omitted.

$$N_2 + 8\,H^+ + 8\,e^- \rightarrow 2\,NH_3 + H_2$$

The process is coupled to the hydrolysis of 16 equivalents of ATP and is accompanied by the co-formation of one molecule of H_2. The conversion of N_2 into ammonia occurs at a cluster called FeMoco, an abbreviation for the iron-molybdenum cofactor. The mechanism proceeds via a series of protonation and reduction steps wherein the FeMoco active site hydrogenates the N_2 substrate.

In free-living diazotrophs, the nitrogenase-generated ammonium is assimilated into glutamate through the glutamine synthetase/glutamate synthase pathway.

The microbial genes required for nitrogen fixation are widely distributed in diverse environments.

Enzymes responsible for nitrogenase action are very susceptible to destruction by oxygen. For this reason, many bacteria cease production of the enzyme in the presence of oxygen. Many nitrogen-fixing organisms exist only in anaerobic conditions, respiring to draw down oxygen levels, or binding the oxygen with a protein such as leghemoglobin.

Microorganisms that Fix Nitrogen

Diazotrophs are a diverse group of prokaryotes that includes cyanobacteria (e.g. the highly significant *Trichodesmium* and *Cyanothece*), green sulfur bacteria, and diazotrophs Azotobacteraceae, rhizobia and *Frankia*.

Cyanobacteria inhabit nearly all illuminated environments on Earth and play key roles in the carbon and nitrogen cycle of the biosphere. In general, cyanobacteria are able to utilize a variety of inorganic and organic sources of combined nitrogen, like nitrate, nitrite, ammonium, urea, or some amino acids. Several cyanobacterial strains are also capable of diazotrophic growth, an ability that may have been present in their last common ancestor in the Archean eon. Nitrogen fixation by cyanobacteria in coral reefs can fix twice the amount of nitrogen as on land—around 1.8 kg of nitrogen is fixed per hectare per day (around 660 kg/ha/year). The colonial marine cyanobacterium *Trichodesmium* is thought to fix nitrogen on such a scale that it accounts for almost half of the nitrogen fixation in marine systems on a global scale.

Root Nodule Symbioses

Legume Family

Plants that contribute to nitrogen fixation include the legume family – Fabaceae – with taxa such as kudzu, clovers, soybeans, alfalfa, lupines, peanuts, and rooibos. They contain symbiotic bacteria called rhizobia within nodules in their root systems, producing nitrogen compounds that help the plant to grow and compete with other plants. When the plant dies, the fixed nitrogen is released, making it available to other plants; this helps to fertilize the soil. The great majority of legumes have this association, but a few genera (e.g., *Styphnolobium*) do not. In many traditional and organic farming practices, fields are rotated through various types of crops, which usually include one consisting mainly or entirely of clover or buckwheat (non-legume family Polygonaceae), often referred to as "green manure".

Inga alley farming relies on the leguminous genus *Inga*, a small tropical, tough-leaved, nitrogen-fixing tree.

Non-Leguminous

Although by far the majority of plants able to form nitrogen-fixing root nodules are in the legume family Fabaceae, there are a few exceptions:

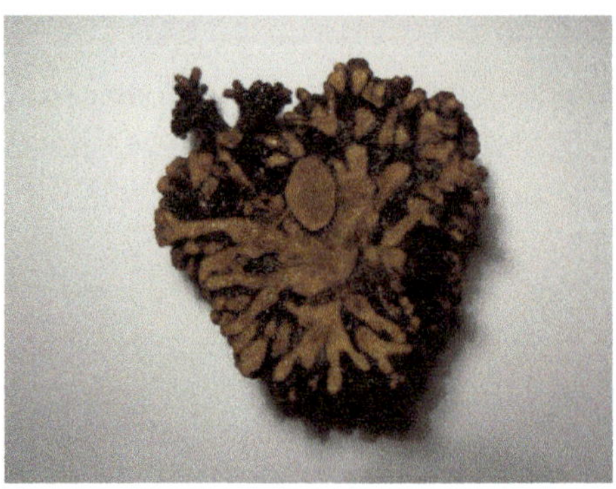

A sectioned alder tree root nodule.

- *Parasponia*, a tropical genus in the Cannabaceae also able to interact with rhizobia and form nitrogen-fixing nodules
- Actinorhizal plants such as alder and bayberry can also form nitrogen-fixing nodules, thanks to a symbiotic association with *Frankia* bacteria. These plants belong to 25 genera distributed among 8 plant families.

The ability to fix nitrogen is far from universally present in these families. For instance, of 122 genera in the Rosaceae, only 4 genera are capable of fixing nitrogen. All these families belong to the orders Cucurbitales, Fagales, and Rosales, which together with the Fabales form a clade of eurosids. In this clade, Fabales were the first lineage to branch off; thus, the ability to fix nitrogen may be plesiomorphic and subsequently lost in most descendants of the original nitrogen-fixing plant; however, it may be that the basic genetic and physiological requirements were present in an incipient state in the last common ancestors of all these plants, but only evolved to full function in some of them:

Family: Genera Betulaceae: *Alnus* (alders) Cannabaceae: *Trema* Casuarinaceae: • Allocasuarina • Casuarina • Ceuthostoma • Gymnostoma	Coriariaceae: *Coriaria* Datiscaceae: *Datisca* Elaeagnaceae: • *Elaeagnus* (silverberries) • *Hippophae* (sea-buckthorns) • *Shepherdia* (buffaloberries)	Myricaceae: • *Comptonia* (sweetfern) • *Morella* • *Myrica* (bayberries)	Rhamnaceae: • Ceanothus • Colletia • Discaria • Kentrothamnus • Retanilla • Talguenea • Trevoa	Rosaceae: • *Cercocarpus* (mountain mahoganies) • *Chamaebatia* (mountain miseries) • *Dryas* • *Purshia/Cowania* (bitterbrushes/cliffroses)

There are also several nitrogen-fixing symbiotic associations that involve cyanobacteria (such as *Nostoc*):

- Some lichens such as *Lobaria* and *Peltigera*
- Mosquito fern (*Azolla* species)
- Cycads
- *Gunnera*

Endosymbiosis in Diatoms

Rhopalodia gibba, a diatom algae, is a eukaryote with cyanobacterial N_2-fixing endosymbiont organelles. The spheroid bodies reside in the cytoplasm of the diatoms and are inseparable from their hosts.

Industrial Nitrogen Fixation

The possibility that atmospheric nitrogen reacts with certain chemicals was first observed by Desfosses in 1828. He observed that mixtures of alkali metal oxides and carbon react at high temperatures with nitrogen. With the use of barium carbonate as starting material the first commercially used process became available in the 1860s developed by Margueritte and Sourdeval. The resulting barium cyanide could be reacted with steam yielding ammonia. In 1898 Adolph Frank and Nikodem Caro decoupled the process and first produced calcium carbide and in a subsequent step reacted it with nitrogen to calcium cyanamide. The Ostwald process for the production of nitric acid was discovered in 1902. Frank-Caro process and Ostwald process dominated the industrial fixation of nitrogen until the discovery of the Haber process in 1909. Prior to 1900, Nikola Tesla also experimented with the industrial fixation of nitrogen "by using currents of extremely high frequency or rate of vibration".

Haber Process

Artificial fertilizer production is now the largest source of human-produced fixed nitrogen in the Earth's ecosystem. Ammonia is a required precursor to fertilizers, explosives, and other products. The most common method is the Haber process. The Haber process requires high pressures (around 200 atm) and high temperatures (at least 400 °C), routine conditions for industrial catalysis. This highly efficient process uses natural gas as a hydrogen source and air as a nitrogen source.

Equipment for a study of nitrogen fixation by alpha rays (Fixed Nitrogen Research Laboratory, 1926)

Much research has been conducted on the discovery of catalysts for nitrogen fixation, often with the goal of reducing the energy required for this conversion. However, such research has thus far failed to even approach the efficiency and ease of the Haber process. Many compounds react with atmospheric nitrogen to give dinitrogen complexes. The first dinitrogen complex to be reported was $Ru(NH_3)_5(N_2)^{2+}$.

Ambient Nitrogen Reduction

Catalytic chemical nitrogen fixation at ambient conditions is an ongoing scientific endeavor. Guided by the example of nitrogenase, this area of homogeneous catalysis is ongoing, with particular emphasis on hydrogenation to give ammonia.

Metallic lithium has long been known for burning in an atmosphere of nitrogen and then converting to lithium nitride. Hydrolysis of the resulting nitride gives ammonia. In a related process, trimethylsilyl chloride, lithium, and nitrogen react in the presence of a catalyst to give tris(trimethylsilyl)amine. Tris(trimethylsilyl)amine can then be used for reaction with α,δ,ω-triketones to give tricyclic pyrroles. Processes involving lithium metal are however of no practical interest since they are noncatalytic and re-reducing the Li^+ ion residue is difficult.

Beginning in the 1960s several homogeneous systems were identified that convert nitrogen to ammonia, sometimes even catalytically but often operating via ill-defined mechanisms. The original discovery is described in an early review:

"Vol'pin and co-workers, using a non-protic Lewis acid, aluminium tribromide, were able to demonstrate the truly catalytic effect of titanium by treating dinitrogen with a mixture of titanium tetrachloride, metallic aluminium, and aluminium tribromide at 50 °C, either in the absence or in the presence of a solvent, e.g. benzene. As much as 200 mol of ammonia per mol of $TiCl_4$ was obtained after hydrolysis...."

The quest for well defined intermediates led to the characterization of many transition metal dinitrogen complexes. Few of these well defined complexes function catalytically, their behavior illuminated likely stages in nitrogen fixation. Most fruitful of all of these early studies focused on $M(N_2)_2(dppe)_2$ (M = Mo, W). For example, double protonation of such low valent complexes gave intermediates with the linkage $M{=}N{-}NH_2$. In 1995, a molybdenum(III) amido complex was discovered that cleaved N_2 to give the corresponding molybdenum(VI) nitride. This and related terminal nitrido complexes have been used to make nitriles.

In 2003 a related molybdenum amido complex was found to catalyze the reduction of N_2. In addition to a source of protons, the catalyst requires a strong reducing agent. However, this catalytic reduction fixates only a few nitrogen molecules. In these systems, like the biological one, hydrogen is provided to the substrate heterolytically, by means of protons and reducing equivalents rather than with H_2 itself.

In 2011 Arashiba et al. reported yet another system with a catalyst again based on molybdenum but with a diphosphorus pincer ligand. Photolytic nitrogen splitting is also considered.

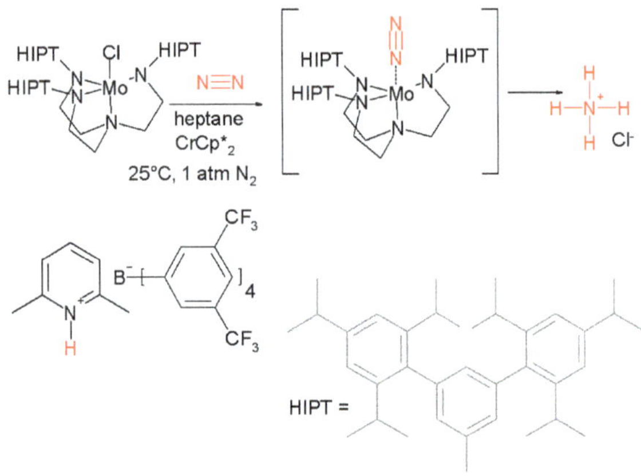

Synthetic nitrogen reduction

Vegetation and Slope Stability

Vegetation and slope stability are interrelated by the ability of the plant life growing on slopes to both promote and hinder the stability of the slope. The relationship is a complex combination of the type of soil, the rainfall regime, the plant species present, the slope aspect, and the steepness of the slope. Knowledge of the underlying slope stability as a function of the soil type, its age, horizon development, compaction, and other impacts is a major underlying aspect of understanding how vegetation can alter the stability of the slope.

There are four major ways in which vegetation influences slope stability: wind throwing, the removal of water, mass of vegetation (surcharge), and mechanical reinforcement of roots.

Wind Throwing

Wind throw is the toppling of a tree due to the force of the wind, this exposes the root plate and adjacent soil beneath the tree and influences slope stability. Wind throw is factor when considering one tree on a slope, however it is of lesser importance when considering general slope stability for a body of trees as the wind forces involved represent a smaller percentage of the potential disturbing forces and the trees which are in the centre of the group will be sheltered by those on the outside.

Removal of Water

Vegetation influences slope stability by removing water through transpiration. Transpiration is the vaporisation of liquid water contained in plant tissue and the vapour removal to the air. Water is drawn up from the roots and transported through the plant up to the leaves.

The major effect of transpiration is the reduction of soil pore water pressures which counteracts the loss of strength which occurs through wetting, this is most readily seen as a loss of moisture around trees. However it is not easy to rely on tree and shrub roots to remove water from slopes and consequently help ensure slope stability. The ability to transpire in wet conditions is severely reduced and therefore any increase in soil strength previously gained in evaporation and transpiration will be lost or significantly reduced, consequently the effects of transpiration cannot be taken into account at these times. However it can be assumed that the chance of slope failure following saturation by storm event or periods of extended rainfall will be lessened as a result of transpiration. Moreover, although changes in moisture content will affect the undrained shear strength, the effective shear stress parameters as commonly used in routine slope stability analysis are not directly influenced by changing moisture content, although the water pressures (suctions) used in the analysis will change.

It is important to note that desiccation cracks can potentially be extended by vegetation in dry weather promoting the deeper penetration of water to a potential slip plane and increased water pressure into the soil during the wet periods. Nevertheless, these cracks will be filled by roots growing deeper into the soil as they follow the path of least resistance.

Studies in Malaysia have shown that there is a significant relationship between root length density, soil water content and ultimately slope stability. Slopes that had high root density (due to dense vegetation on the surface) were less likely to undergo slope failure. This is because a high root length density results in low soil water content which in turn results in an increase in shear strength and a decrease in soil permeability. It is suggested that root length density and soil water level could be used as indicators of slope stability and possibly could be used to predict future slope failure.

Transpiration is accentuated when the vegetation has an extensive root system and rapid transpiration continues throughout winter.

The removal of water is also affected by the shading provided by vegetation. Shading helps prevent the desiccation of the soils which results in shrinkage and cracking allowing the deep penetration of rain water. Plants need to have a high leaf to root ratio and have the ability to persist through hot summer months in order to provide effective shading of the soils.

The Mass of Vegetation

The mass of vegetation is only likely to have an influence on slope stability when larger trees are growing on the slope. A tree of 30 -50m height is likely to have a loading of approximately 100 -150 kN/m2. The larger trees should be planted at the toe of the slope with a potential rotational failure as this could increase the factor of safety by 10%. However, if the tree is planted at the top of the slope this could reduce the factor of safety by 10%. A suggested design vegetation envelope which shows which type of plant should be grown at each level of the slope.

Each slope stability situation should be considered independently for the vegetation involved. It is important to remember that transpiration will reduce the weight of the slope as moisture is lost. This can be significant on slopes of marginal stability.

If larger trees are removed from the toe area of a slope there will be both a reduction in soil strength due to the loss of evapotranspiration effects and a reduction in applied loading which may result in temporary suctions in clay soils which could lead to softening as the available water is drawn in to compensate for the suction forces. This is similar to the recognised softening of overconsolidated clays due to the relaxation of overburden pressures when placed in the top layers of an embankment from deep cutting.

Mechanical Reinforcement of Roots

Roots reinforce the soil through growing across failure planes, root columns acting as piles, and through limiting surface erosion.

Root Growth Across Failure Planes

When roots grow across the plane of potential failure there is an increase in shear strength by binding particles. The roots anchor the unstable surficial soil into the deeper stable layers or bedrock. This most readily occurs when there is rapid deep growth (1.5m deep) of roots which last for more than two years. However it is important to note that the strength exerted by roots generally only extends down to 1m while most failures occur between 1.2 – 1.5m soil depth.

Root Reinforcement Model

The root reinforced earth root model is the result of the root elongation across a potential slip plane which produces a tensile root force which is transferred to the soil by cohesive and frictional contacts between the root and the soil.

Tensile Root Strength Contribution and Pull Out Resistance

The pull out resistance of a root is the measured resistance of root structure to be pulled out of the ground and is likely to be only a little less than the measured tensile strength of the root which is the roots resistance to breaking as measured in the laboratory. In the cases where there is no pull out data available the tensile strength data maybe used as a rough guide to the maximum pull out resistance available.

The tensile root strength of a range of diameters over a range of species has been tested in the laboratory and has been found to be approximately 5 – 60MN/m2. In order for the root to actually enhance slope stability the root must have sufficient embedment and adhesion with the soil. The way that roots interact with the soil is intricate but for engineering purposes the available force contributions may be measured with in situ pull out tests.

Root Morphology and Modes of Failure

The root length and the type of root branching effects the way that root failure occurs Three different modes of failure have been identified in hawthorn roots which relate to the root soil relationship which is shown in the shape of the roots and the shape of the failure curve. Roots which have no branches tend to fail in tension and pull straight out of the ground with minimal resistance. Roots which have multiple branches generally fail in stages as each branch breaks inside the soil. These roots can then separated into two different groups; 1) those that initially reach their maximum peak force and then maintain a high force that progressively decreases as the root branches fail after significant strain and 2) those that break with increasingly applied force. In a number of tests considerable adhesion between a segment of the root and the soil can be measured prior to the root eventually slipping out of the soil mass.

Type A Failure

Roots that do not have branches generally fail in tension and pull straight out of the ground with only minimal resistance. The root reaches its maximum pullout resistance then rapidly fails at a weak point. The root easily slips out of the soil due to the gradual tapering (progressive decrease in root diameter along its length) which means that as the root is pulled out it is moving through a space that is larger than its diameter which consequently has no further bonds or interaction with the surrounding soil.

Type B Failure

Type B failure occurs when branched roots initially reach their maximum peak resistance then sustain a high resistance which slowly reduces as the branches of the roots fail after significant strain. In some tests considerable adhesion between a section of the root and the soil mass can be measured before the root eventually slips out. Forked roots require a greater force to be pulled out as the cavity above the fork is thinner than the root which is trying to move through the cavity, this can then result in deformation of the soil as the root moves through the soil.

Type C Failure

Roots that have multiple branches or forked branches also can undergo tensile failure but predominantly fail in stages as each branch breaks within the soil. These roots break with increasingly applied force in stages in the form of stepped peaks corresponding to the progressive breaking of roots of greater diameters. The root progressively releases its bonds with the soil until final tensile failure.

In some cases when the root has a sinusoidal shape with many small rootlets along its length the root reaches its maximum pull out resistance on straightening and then breaks at the weakest point, however at this point the root is not pulled out of the soil as it adheres and interacts with the soil producing a residual strength. If pulling was stopped at this point, the root would give increased strength to the soil. However, if the root is completely pulled out of the ground then there is no further interaction with soil and therefore no increase in soil strength is provided.

Factors which Affect Root Pull Out Resistance

Studies have shown that the pull out resistance of hawthorn and oak roots are affected by intra species differences, inter-species variations and root size (diameter) in a similar as way as root tensile strength varies (as measured in the laboratory). In the pull out test the applied force acting on the root acts across a larger root area, which involves multiple branches, longer lengths) than the short (approximately 150mm) length of root used in tensile strength tests. In pull out test the root is likely to fail at weak points such as branching points, nodes or damaged areas.

The studies also showed that there is a positive correlation between maximum root pull out resistance and root diameter for hawthorn and oat root. Smaller diameter roots had a lower pull out resistance or breaking force than the larger diameter roots.

Root Columns Acting as Piles

Trees and root columns can prevent shallow mass movement through acting as piles when there is buttressing and soil arching through a woody deep root system which has multiple sinker roots with embedded stems and laterals.

Limitation of Surface Erosion

Vegetation can also be used to control water erosion by limiting surface processes such as sheet wash and overland flow. Vegetation can provide a considerable contribution to the stability of slope through enhancing soil cohesion. This cohesion is dependent upon the morphological characteristics of root systems and the tensile strength of single roots.

There is considerable evidence of fine roots resisting surface erosion. The role of fine roots in general slope stability is not fully understood. It is thought that the fine roots help keep the surface soil together and prevent surface erosion. The fine root network may have an apparent enhanced cohesion which is comparable to geosynthetic mesh elements. The limitation of surface erosion processes is particularly apparent in areas of shrub and grass where the fine root distribution is consistent and clearly defined, however cohesion is generally limited to the top 1m of soil.

Tilth

Soil tilth is its physical condition, especially in relation to its suitability for planting or growing a crop. Factors that determine tilth include the formation and stability of aggregated soil particles, moisture content, degree of aeration, rate of water infiltration and drainage. Tilth can change rapidly, depending on environmental factors such as changes in moisture, tillage and soil amendments.

Soil with good tilth has large pore spaces for air infiltration and water movement. Roots only grow where the soil tilth allows for adequate levels of soil oxygen. Such soil also holds a reasonable supply of water and nutrients.

Tillage, organic matter amendments, fertilization and irrigation can each improve tilth, but when used excessively, can have the opposite effect. Crop rotation and cover crops can positively impact tilth. A combined approach can produce the greatest improvement.

Aggregation

Aggregation is positively associated with tilth. With finer-textured soils, aggregates may in turn be made up of smaller aggregates. Aggregation implies substantial pores between individual aggregates.

Aggregation is important in the subsoil, the layer below tillage. Such aggregates involve larger (2- to 6-inch) blocks of soil that are more angular and not as distinctive. These aggregates are less impacted by biological activity than the tillage layer. Subsurface aggregates are important for root growth deep into the profile. Deep roots allow greater access to moisture, which helps in drought periods. Subsoil aggregates can also be compacted, mainly by heavy equipment on wet soil. Another significant source of subsoil compaction is the practice of plowing with tractor wheels in the open furrow.

Pore Size

Soil that is well aggregated has a range of pore sizes. Each pore size plays a role in soil's physical functioning. Large pores drain rapidly and are needed for good air exchange during wet periods, preventing oxygen deficiency that can drown plants and increase pest problems. Denitrification by conversion of nitrogen to gaseous forms is increased in oxygen-deficient wet soil. In degraded soil large pores are compressed into small ones.

Small pores are critical for water retention and help a crop endure dry periods with minimal yield loss.

Management

Soil tilth can be obtained through mechanical and biological manipulation.

Tillage

Mechanical soil cultivation practices, including primary tillage (mold-board or chisel plowing) followed by secondary tillage (disking, harrowing, etc.), break up and aerate soil. When soils become degraded and compacted, such tillage practices are often deemed necessary. The tilth created by tillage, however, tends to be *unstable*, because the aggregation is obtained through the physical manipulation of the soil, which is short lived, especially after years of intensive tillage. Aggregates in such soils readily dissolve during subsequent rains, causing the soil to settle and become dense and hard, requiring further tillage.

Biological

The preferred scenario for good tilth is as the result of natural soil-building processes, provided by the activity of plant roots, microorganisms, earthworms and other beneficial organisms. Such stable aggregates break apart during tillage/planting and readily provide good tilth. Stable aggregates are held together by organic bonds that resist breakdown during soil saturation. These organic materials are themselves subject to biological degradation, requiring active amendments with organic material, and minimal mechanical tillage.

Rotation

Crop rotation can help restore tilth in compacted soils. Two processes contribute to this gain. First, accelerated organic matter decomposition from tillage ends under the sod crop. Another way to achieve this is via no-till farming. Second, grass and legume sods develop extensive root systems that continually grow and die off. The dead roots supply a source of active organic matter, which feeds soil organisms that create aggregation. Beneficial organisms need continual supplies of organic matter to sustain themselves and they deposit the digested materials on soil aggregates and thereby stabilize them. Also, the living roots and symbiotic microorganisms (for example, mycorrhizal fungi) can exude organic materials that nourish soil organisms and help with aggregation. Grass and legume sod crops therefore return more organic matter to the soil than most other crops.

Some annual rotation crops such as buckwheat also have dense, fibrous, root systems and can improve tilth. Crop mixtures with different rooting systems can be beneficial. For example, red clover seeded into winter wheat provide additional roots and a more protein-rich organic matter.

Other rotation crops are more valuable for improving subsoils. Perennial crops such as alfalfa have strong, deep, penetrating tap roots that can push through hard layers, especially during wet periods when the soil is soft. These deep roots establish pathways for water and future plant roots, and produce organic matter.

Crops rotation can extend the period of active growth compared to conventional row crops, leaving more organic material behind. For example, in a corn-soybean rotation, active growth occurs 32 percent of the time, while a dry bean–winter wheat–corn rotation is active 72 percent. Crops such as rye, wheat, oat, barley, pea and cool-season grasses grow actively in the late fall and early spring when other crops are inactive. They are beneficial both as rotation and cover crops, although intensive tillage can negate their effects.

Soil Types

The soil management practices required to maintain soil tilth are a function of the type of soil. Sandy and gravelly soils are naturally deficient in small pores and are therefore drought prone, whereas loams and clays can retain and thus supply crops with more water.

Coarse-Textured, Sandy Soils

Sandy soil has lower capacity to hold water and nutrients. Water is applied more frequently in smaller amounts to avoid it leaching and carrying nutrients below the root zone. Routine application of organic matter increases sandy soil's ability to hold water and nutrients by 10 times or more.

Fine-Textured, Clay Soils

Clay soils lack large pores, restricting both water and air movement. During irrigation or rain events, the limited large pore space in fine-textured soils quickly fills with water, reducing soil oxygen levels. In addition to routine application of organic matter, microorganisms and earthworms perform a crucial assist to soil tilth. As microorganisms decompose the organic matter, soil particles bind together into larger aggregates, increasing large pore space. Clay soils are more subject to soil compaction, which reduces large pore spaces.

Gravelly and Decomposed Granite Soils

Such soils natively have little tilth, especially once they have been disturbed. Adding organic matter up to 25% by volume can help compensate. For example, if tilling to a depth of eight inches, add two inches of organic materials.

References

- McCarthy, David F. (2007). Essentials of Soil Mechanics and Foundations. Upper Saddle River, NJ: Pearson Prentice Hall. p. 595. ISBN 0-13-114560-6.
- Groves, C.P. (2005). Wilson, D.E.; Reeder, D.M., eds. Mammal Species of the World: A Taxonomic and Geographic Reference (3rd ed.). Baltimore: Johns Hopkins University Press. OCLC 62265494. ISBN 0-801-88221-4.
- McHenry, H.M (2009). "Human Evolution". In Michael Ruse & Joseph Travis. Evolution: The First Four Billion Years. Cambridge, Massachusetts: The Belknap Press of Harvard University

5

Soil Pollution and Erosion: An Integrated Study

The harm caused by the presence of human made chemicals and industrial waste in the soil is soil pollution. The cause of this pollution is industrial activities, agricultural chemicals and improper disposal of waste. The following content explains to the reader soil retrogression and degradation.

Soil Contamination

Soil contamination or soil pollution as part of land degradation is caused by the presence of xenobiotic (human-made) chemicals or other alteration in the natural soil environment. It is typically caused by industrial activity, agricultural chemicals, or improper disposal of waste. The most common chemicals involved are petroleum hydrocarbons, polynuclear aromatic hydrocarbons (such as naphthalene and benzo(a)pyrene), solvents, pesticides, lead, and other heavy metals. Contamination is correlated with the degree of industrialization and intensity of chemical usage.

Excavation showing soil contamination at a disused gasworks.

The concern over soil contamination stems primarily from health risks, from direct contact with the contaminated soil, vapors from the contaminants, and from secondary contamination of water supplies within and underlying the soil. Mapping of contaminated soil sites and the resulting cleanup are time consuming and expensive tasks, requiring extensive amounts of geology, hydrology, chemistry, computer modeling skills, and GIS in Environmental Contamination, as well as an appreciation of the history of industrial chemistry.

In North America and Western Europe the extent of contaminated land is best known, with many of countries in these areas having a legal framework to identify and deal with this environmental problem. Developing countries tend to be less tightly regulated despite some of them having undergone significant industrialization.

Causes

Soil pollution can be caused by the following (non-exhaustive list!):

- Oil drilling
- Mining and activities by other heavy industries
- Accidental spills as may happen during deforestation activities, etc.
- Corrosion of underground storage tanks (including piping used to transmit the contents)
- Acid rain (in turn caused by air pollution)
- Intensive farming
- Agrochemicals, such as pesticides, herbicides and fertilizers
- Industrial accidents
- Road debris
- Drainage of contaminated surface water into the soil
- Waste disposal
    - Oil and fuel dumping
    - Nuclear wastes
    - Direct discharge of industrial wastes to the soil
    - Landfill and illegal dumping
    - coal ash
    - Electronic waste
    - ammunitions and agents of war

The most common chemicals involved are petroleum hydrocarbons, solvents, pesticides, lead, and other heavy metals.

In a wider sense, genetically modified plants (GMP) can count as a risk factor for soils, because of their potential to affect the soil fauna. Any activity that leads to other forms of soil degradation (erosion, compaction, etc.) may indirectly worsen the contamination effects in that soil remediation becomes more tedious.

Historical deposition of coal ash used for residential, commercial, and industrial heating, as well as for industrial processes such as ore smelting, were a common source of contamination in areas that were industrialized before about 1960. Coal naturally concentrates lead and zinc during its formation, as well as other heavy metals to a lesser degree. When the coal is burned, most of these metals become concentrated in the ash (the principal exception being mercury). Coal ash and slag may contain sufficient lead to qualify as a "characteristic hazardous waste", defined in the USA as containing more than 5 mg/l of extractable lead using the TCLP procedure. In addition to lead, coal ash typically contains variable but significant concentrations of polynuclear aromatic hydrocarbons (PAHs; e.g., benzo(a) anthracene, benzo(b)fluoranthene, benzo(k)fluoranthene, benzo(a)pyrene, indeno(cd) pyrene, phenanthrene, anthracene, and others). These PAHs are known human carcinogens and the acceptable concentrations of them in soil are typically around 1 mg/kg. Coal ash and slag can be recognised by the presence of off-white grains in soil, gray heterogeneous soil, or (coal slag) bubbly, vesicular pebble-sized grains.

Treated sewage sludge, known in the industry as biosolids, has become controversial as a "fertilizer". As it is the byproduct of sewage treatment, it generally contains more contaminants such as organisms, pesticides, and heavy metals than other soil.

In the European Union, the Urban Waste Water Treatment Directive allows sewage sludge to be sprayed onto land. The volume is expected to double to 185,000 tons of dry solids in 2005. This has good agricultural properties due to the high nitrogen and phosphate content. In 1990/1991, 13% wet weight was sprayed onto 0.13% of the land; however, this is expected to rise 15 fold by 2005. Advocates say there is a need to control this so that pathogenic microorganisms do not get into water courses and to ensure that there is no accumulation of heavy metals in the top soil.

Pesticides and Herbicides

A pesticide is a substance or mixture of substances used to kill a pest. A pesticide may be a chemical substance, biological agent (such as a virus or bacteria), antimicrobial, disinfectant or device used against any pest. Pests include insects, plant pathogens, weeds, mollusks, birds, mammals, fish, nematodes (roundworms) and microbes that compete with humans for food, destroy property, spread or are a vector for disease or cause a nuisance. Although there are benefits to the use of pesticides, there are also drawbacks, such as potential toxicity to humans and other organisms.

Herbicides are used to kill weeds, especially on pavements and railways. They are similar to auxins and most are biodegradable by soil bacteria. However, one group derived from trinitrotoluene (2:4 D and 2:4:5 T) have the impurity dioxin, which is very toxic and causes fatality even in low concentrations. Another herbicide is Paraquat. It is highly toxic but it rapidly degrades in soil due to the action of bacteria and does not kill soil fauna.

Insecticides are used to rid farms of pests which damage crops. The insects damage not only standing crops but also stored ones and in the tropics it is reckoned that one third of the total production is lost during food storage. As with fungicides, the first insecticides used in the nineteenth century were inorganic e.g. Paris Green and other compounds of arsenic. Nicotine has also been used since the late eighteenth century.

There are now two main groups of synthetic insecticides -

1. Organochlorines include DDT, Aldrin, Dieldrin and BHC. They are cheap to produce, potent and persistent. DDT was used on a massive scale from the 1930s, with a peak of 72,000 tonnes used 1970. Then usage fell as the harmful environmental effects were realized. It was found worldwide in fish and birds and was even discovered in the snow in the Antarctic. It is only slightly soluble in water but is very soluble in the bloodstream. It affects the nervous and endocrine systems and causes the eggshells of birds to lack calcium causing them to be easily breakable. It is thought to be responsible for the decline of the numbers of birds of prey like ospreys and peregrine falcons in the 1950s - they are now recovering. As well as increased concentration via the food chain, it is known to enter via permeable membranes, so fish get it through their gills. As it has low water solubility, it tends to stay at the water surface, so organisms that live there are most affected. DDT found in fish that formed part of the human food chain caused concern, but the levels found in the liver, kidney and brain tissues was less than 1 ppm and in fat was 10 ppm, which was below the level likely to cause harm. However, DDT was banned in the UK and the United States to stop the further buildup of it in the food chain. U.S. manufacturers continued to sell DDT to developing countries, who could not afford the expensive replacement chemicals and who did not have such stringent regulations governing the use of pesticides..

2. Organophosphates, e.g. parathion, methyl parathion and about 40 other insecticides are available nationally. Parathion is highly toxic, methyl-parathion is less so and Malathion is generally considered safe as it has low toxicity and is rapidly broken down in the mammalian liver. This group works by preventing normal nerve transmission as cholinesterase is prevented from breaking down the transmitter substance acetylcholine, resulting in uncontrolled muscle movements.

Agents of War

The disposal of munitions, and a lack of care in manufacture of munitions caused by the urgency of production, can contaminate soil for extended periods. There is little published evidence on this type of contamination largely because of restrictions placed by Governments of many countries on the publication of material related to war effort. However, mustard gas stored during World War II has contaminated some sites for up to 50 years and the testing of Anthrax as a potential biological weapon contaminated the whole island of Gruinard

Health Effects

Contaminated or polluted soil directly affects human health through direct contact with soil or via inhalation of soil contaminants which have vaporized; potentially greater threats are posed by the infiltration of soil contamination into groundwater aquifers used for human consumption, sometimes in areas apparently far removed from any apparent source of above ground contamination. This tends to result in the development of pollution-related diseases.

Health consequences from exposure to soil contamination vary greatly depending on pollutant type, pathway of attack and vulnerability of the exposed population. Chronic exposure to chromium, lead and other metals, petroleum, solvents, and many pesticide and herbicide formulations can be carcinogenic, can cause congenital disorders, or can cause other chronic health conditions. Industrial or man-made concentrations of naturally occurring substances, such as nitrate and ammonia associated with livestock manure from agricultural operations, have also been identified as health hazards in soil and groundwater.

Chronic exposure to benzene at sufficient concentrations is known to be associated with higher incidence of leukemia. Mercury and cyclodienes are known to induce higher incidences of kidney damage and some irreversible diseases. PCBs and cyclodienes are linked to liver toxicity. Organophosphates and carbomates can induce a chain of responses leading to neuromuscular blockage. Many chlorinated solvents induce liver changes, kidney changes and depression of the central nervous system. There is an entire spectrum of further health effects such as headache, nausea, fatigue, eye irritation and skin rash for the above cited and other chemicals. At sufficient dosages a large number of soil contaminants can cause death by exposure via direct contact, inhalation or ingestion of contaminants in groundwater contaminated through soil.

The Scottish Government has commissioned the Institute of Occupational Medicine to undertake a review of methods to assess risk to human health from contaminated land. The overall aim of the project is to work up guidance that should be useful to Scottish Local Authorities in assessing whether sites represent a significant possibility of significant harm (SPOSH) to human health. It is envisaged that the output of the project will be a short document providing high level guidance on health risk assessment with reference to existing published guidance and methodologies that have been identified as being particularly relevant and helpful. The project will examine how policy guidelines have been developed for determining the acceptability of risks to human health and propose an approach for assessing what constitutes unacceptable risk in line with the criteria for SPOSH as defined in the legislation and the Scottish Statutory Guidance.

Ecosystem Effects

Not unexpectedly, soil contaminants can have significant deleterious consequences for ecosystems. There are radical soil chemistry changes which can arise from the presence of many hazardous chemicals even at low concentration of the contaminant species. These changes can manifest in the alteration of metabolism of endemic microorganisms and arthropods resident in a given soil environment. The result can be virtual eradication of some of the primary food chain, which in turn could have major consequences for predator or consumer species. Even if the chemical effect on lower life forms is small, the lower pyramid levels of the food chain may ingest alien chemicals, which normally become more concentrated for each consuming rung of the food chain. Many of these effects are now well known, such as the concentration of persistent DDT materials for avian consumers, leading to weakening of egg shells, increased chick mortality and potential extinction of species.

Effects occur to agricultural lands which have certain types of soil contamination. Contaminants typically alter plant metabolism, often causing a reduction in crop yields. This has a secondary effect upon soil conservation, since the languishing crops cannot shield the Earth's soil from erosion. Some of these chemical contaminants have long half-lives and in other cases derivative chemicals are formed from decay of primary soil contaminants.

Cleanup Options

Cleanup or environmental remediation is analyzed by environmental scientists who utilize field measurement of soil chemicals and also apply computer models (GIS in Environmental Contamination) for analyzing transport and fate of soil chemicals. Various technologies have been developed for remediation of oil-contaminated soil/ sediments There are several principal strategies for remediation:

- Excavate soil and take it to a disposal site away from ready pathways for human or sensitive ecosystem contact. This technique also applies to dredging of bay muds containing toxins.
- Aeration of soils at the contaminated site (with attendant risk of creating air pollution)
- Thermal remediation by introduction of heat to raise subsurface temperatures sufficiently high to volatize chemical contaminants out of the soil for vapour extraction. Technologies include ISTD, electrical resistance heating (ERH), and ET-DSP^{tm}.
- Bioremediation, involving microbial digestion of certain organic chemicals. Techniques used in bioremediation include landfarming, biostimulation and

bioaugmentating soil biota with commercially available microflora.

- Extraction of groundwater or soil vapor with an active electromechanical system, with subsequent stripping of the contaminants from the extract.
- Containment of the soil contaminants (such as by capping or paving over in place).
- Phytoremediation, or using plants (such as willow) to extract heavy metals
- Mycoremediation, or using fungus to metabolize contaminants and accumulate heavy metals.

By Country

Various national standards for concentrations of particular contaminants include the United States EPA Region 9 Preliminary Remediation Goals (U.S. PRGs), the U.S. EPA Region 3 Risk Based Concentrations (U.S. EPA RBCs) and National Environment Protection Council of Australia Guideline on Investigation Levels in Soil and Groundwater.

People's Republic of China

The immense and sustained growth of the People's Republic of China since the 1970s has exacted a price from the land in increased soil pollution. The State Environmental Protection Administration believes it to be a threat to the environment, to food safety and to sustainable agriculture. According to a scientific sampling, 150 million mi (100,000 square kilometers) of China's cultivated land have been polluted, with contaminated water being used to irrigate a further 31.5 million mi (21,670 km^2.) and another 2 million mi (1,300 square kilometers) covered or destroyed by solid waste. In total, the area accounts for one-tenth of China's cultivatable land, and is mostly in economically developed areas. An estimated 12 million tonnes of grain are contaminated by heavy metals every year, causing direct losses of 20 billion yuan (US$2.57 billion).

United Kingdom

Generic guidance commonly used in the United Kingdom are the Soil Guideline Values published by the Department for Environment, Food and Rural Affairs (DEFRA) and the Environment Agency. These are screening values that demonstrate the minimal acceptable level of a substance. Above this there can be no assurances in terms of significant risk of harm to human health. These have been derived using the Contaminated Land Exposure Assessment Model (CLEA UK). Certain input parameters such as Health Criteria Values, age and land use are fed into CLEA UK to obtain a probabilistic output.

Guidance by the Inter Departmental Committee for the Redevelopment of Contaminated Land (ICRCL) has been formally withdrawn by DEFRA, for use as a prescriptive document to determine the potential need for remediation or further assessment.

The CLEA model published by DEFRA and the Environment Agency (EA) in March 2002 sets a framework for the appropriate assessment of risks to human health from contaminated land, as required by Part IIA of the Environmental Protection Act 1990. As part of this framework, generic Soil Guideline Values (SGVs) have currently been derived for ten contaminants to be used as "intervention values". These values should not be considered as remedial targets but values above which further detailed assessment should be considered.

Three sets of CLEA SGVs have been produced for three different land uses, namely

- residential (with and without plant uptake)
- allotments
- commercial/industrial

It is intended that the SGVs replace the former ICRCL values. It should be noted that the CLEA SGVs relate to assessing chronic (long term) risks to human health and do not apply to the protection of ground workers during construction, or other potential receptors such as groundwater, buildings, plants or other ecosystems. The CLEA SGVs are not directly applicable to a site completely covered in hardstanding, as there is no direct exposure route to contaminated soils.

To date, the first ten of fifty-five contaminant SGVs have been published, for the following: arsenic, cadmium, chromium, lead, inorganic mercury, nickel, selenium ethyl benzene, phenol and toluene. Draft SGVs for benzene, naphthalene and xylene have been produced but their publication is on hold. Toxicological data (Tox) has been published for each of these contaminants as well as for benzo[a]pyrene, benzene, dioxins, furans and dioxin-like PCBs, naphthalene, vinyl chloride, 1,1,2,2 tetrachloroethane and 1,1,1,2 tetrachloroethane, 1,1,1 trichloroethane, tetrachloroethene, carbon tetrachloride, 1,2-dichloroethane, trichloroethene and xylene. The SGVs for ethyl benzene, phenol and toluene are dependent on the soil organic matter (SOM) content (which can be calculated from the total organic carbon (TOC) content). As an initial screen the SGVs for 1% SOM are considered to be appropriate.

Canada

India

In March 2009, the issue of Uranium poisoning in Punjab attracted press coverage. It was alleged to be caused by fly ash ponds of thermal power stations, which reportedly lead to severe birth defects in children in the Faridkot and Bhatinda districts of Punjab. The news reports claimed the uranium levels were more than 60 times the maximum safe limit. In 2012, the Government of India confirmed that the ground water in Malwa belt of Punjab has uranium metal that is 50% above the trace limits set by the United Nations' World Health Organization. Scientific studies, based on over 1000 samples from various sampling points, could not trace the source to fly ash and any sources from thermal power plants or industry as originally alleged. The study also revealed that the uranium concentration in ground water of Malwa district is not 60 times the WHO limits, but only 50% above the WHO limit in 3 locations. This highest concentration found in samples was less than those found naturally in ground waters currently used for human purposes elsewhere, such as Finland. Research is underway to identify natural or other sources for the uranium.

Causes of Soil Erosion

Hydraulic Action

Hydraulic action is erosion that occurs when the motion of water against a rock surface produces mechanical weathering. Most generally, it is the ability of moving water (flowing or waves) to dislodge and transport rock particles. Within this rubric are a number of specific erosional processes, including abrasion, attrition, corrasion, saltation, and scouring. Hydraulic action is distinguished from other types of water facilitated erosion, such as *static erosion* where water leaches salts and floats off organic material from unconsolidated sediments, and from *chemical erosion* more often called chemical weathering. It is a mechanical process, in which the moving water current flows against the banks and bed of a river, thereby removing rock particles.

A primary example of hydraulic action is a wave striking a cliff face which compresses the air in cracks of the rocks. This exerts pressure on the surrounding rock which can progressively crack, break, splinter and detach rock particles. This is followed by the decompression of the air as the wave retreats which can occur suddenly with explosive force which additionally weakens the rock. Cracks are gradually widened so each wave compresses more air, increasing the explosive force of its release. Thus, the effect intensifies in a 'positive feedback' system. Over time, as the cracks may grow they sometimes form a sea cave. The broken pieces that fall off produce two additional types of erosion, abrasion (sandpapering) and attrition. In corrasion, the newly formed chunks are thrown against the rock face. Attrition is a similar effect caused by eroded particles after they fall to the sea bed where they are subjected to further wave action. In coastal areas wave hydraulic action is often the most important form of erosion.

Similarly, where hydraulic action is strong enough to loosen sediment along a stream bed and its banks; this will take rocks and particles from the banks and bed of the stream and add this to the stream's load. This process is the result of friction between the moving water and the static stream bed and banks. This friction increases with the speed of the water and the roughness of the bed. Once loosened the smaller particles are actually held in suspension by the force of the flowing water, these suspended particles can scour the sides and bottom of the stream. The scouring action produces distinctive markings on streams beds such as ripple marks, fluting, and crescent marks. The larger particles and even large rocks are *scooted* (dragged) along the bottom in a process known as *traction* which causes attrition, and are often "bounced" along in a process known as saltation where the force of the water temporarily lifts the rock particle which then crashes back into the bed dislodging other particles.

Hydraulic action also occurs as a stream tumbles over a waterfall to crash onto the rocks below. It usually leads to the formation of a plunge pool below the waterfall due in part to corrosion from the stream's load, but more to a scouring action as vortices form in the water as it escapes downstream. Hydraulic action can also cause the breakdown of river banks since there are water bubbles which enter the banks and collapse them when they expand.

Aeolian Processes

Wind erosion of soil at the foot of Chimborazo, Ecuador.

Aeolian processes, also spelled eolian or æolian, pertain to wind activity in the study of geology and weather and specifically to the wind's ability to shape the surface of the Earth (or other planets). Winds may erode, transport, and deposit materials and are effective agents in regions with sparse vegetation, a lack of soil moisture and a large supply of unconsolidated sediments. Although water is a much more powerful eroding force than wind, aeolian processes are important in arid environments such as deserts.

Rock carved by drifting sand below Fortification Rock in Arizona (Photo by Timothy H. O'Sullivan, USGS, 1871)

The term is derived from the name of the Greek god Aeolus, the keeper of the winds.

Wind Erosion

A rock sculpted by wind erosion in the Altiplano region of Bolivia

Wind erodes the Earth's surface by deflation (the removal of loose, fine-grained particles by the turbulent action of the wind) and by abrasion (the wearing down of surfaces by the grinding action and sandblasting of windborne particles).

Sand blowing off a crest in the Kelso Dunes of the Mojave Desert, California.

Regions which experience intense and sustained erosion are called deflation zones. Most aeolian deflation zones are composed of desert pavement, a sheet-like surface of rock fragments that remains after wind and water have removed the fine particles. Almost half of Earth's desert surfaces are stony deflation zones. The rock mantle in desert pavements protects the underlying material from deflation.

Wind-carved alcove in the Navajo Sandstone near Moab, Utah

A dark, shiny stain, called desert varnish or rock varnish, is often found on the surfaces of some desert rocks that have been exposed at the surface for a long period of time. Manganese, iron oxides, hydroxides, and clay minerals form most varnishes and provide the shine.

Deflation basins, called blowouts, are hollows formed by the removal of particles by wind. Blowouts are generally small, but may be up to several kilometers in diameter.

Wind-driven grains abrade landforms. In parts of Antarctica wind-blown snowflakes that are technically sediments have also caused abrasion of exposed rocks. Grinding by particles carried in the wind creates grooves or small depressions. Ventifacts are rocks which have been cut, and sometimes polished, by the abrasive action of wind.

Sculpted landforms, called yardangs, are up to tens of meters high and kilometers long and are forms that have been streamlined by desert winds. The famous Great Sphinx of Giza in Egypt may be a modified yardang.

Transport

Particles are transported by winds through suspension, saltation (skipping or bouncing) and creeping (rolling or sliding) along the ground.

Dust storm approaching Spearman, Texas April 14, 1935.

Small particles may be held in the atmosphere in suspension. Upward currents of air support the weight of suspended particles and hold them indefinitely in the surrounding air. Typical winds near Earth's surface suspend particles less than 0.2 millimeters in diameter and scatter them aloft as dust or haze.

Saltation is downwind movement of particles in a series of jumps or skips. Saltation normally lifts sand-size particles no more than one centimeter above the ground and proceeds at one-half to one-third the speed of the wind. A saltating grain may hit other grains that jump up to continue the saltation. The grain may also hit larger grains that are too heavy to hop, but that slowly creep forward as they are pushed by saltating grains. Surface creep accounts for as much as 25 percent of grain movement in a desert.

Aeolian turbidity currents are better known as dust storms. Air over deserts is cooled significantly when rain passes through it. This cooler and denser air sinks toward the desert surface. When it reaches the ground, the air is deflected forward and sweeps up surface debris in its turbulence as a dust storm.

Dust storm in Amarillo, Texas. FSA photo by Arthur Rothstein (1936)

Crops, people, villages, and possibly even climates are affected by dust storms. Some dust storms are intercontinental, a few may circle the globe, and occasionally they may engulf entire planets. When the Mariner 9 spacecraft entered its orbit around Mars in 1971, a dust storm lasting one month covered the entire planet, thus delaying the task of photo-mapping the planet's surface.

A massive sand storm cloud is about to envelop a military camp as it rolls over Al Asad, Iraq, just before nightfall on April 27, 2005.

Most of the dust carried by dust storms is in the form of silt-size particles. Deposits of this windblown silt are known as loess. The thickest known deposit of loess, 335 meters, is on the Loess Plateau in China. This very same Asian dust is blown for thousands of miles, forming deep beds in places as far away as Hawaii. In Europe and in the Americas, accumulations of loess are generally from 20 to 30 meters thick.

Aeolian transport from deserts plays an important role in ecosystems globally, e.g. by transport of minerals from the Sahara to Amazonia. Saharan dust is also responsible for forming red clay soils in southern Europe. Aeolian processes are affected by human activity, such as the use of 4x4 vehicles.

Small whirlwinds, called dust devils, are common in arid lands and are thought to be related to very intense local heating of the air that results in instabilities of the air mass. Dust devils may be as much as one kilometer high.

Deposition

Wind-deposited materials hold clues to past as well as to present wind directions and intensities. These features help us understand the present climate and the forces that molded it. Wind-deposited sand bodies occur as sand sheets, ripples, and dunes.

Cross-bedding of sandstone near Mount Carmel road, Zion National Park, indicating wind action and sand dune formation prior to formation of rock (NPS photo by George A. Grant, 1929)

Sand sheets are flat, gently undulating sandy plots of sand surfaced by grains that may be too large for saltation. They form approximately 40 percent of aeolian depositional surfaces. The Selima Sand Sheet in the eastern Sahara Desert, which occupies 60,000 square kilometers in southern Egypt and northern Sudan, is one of the Earth's largest sand sheets. The Selima is absolutely flat in a few places; in others, active dunes move over its surface.

Mesquite Flat Dunes in Death Valley looking toward the Cottonwood Mountains from the north west arm of Star Dune (2003)

Wind blowing on a sand surface ripples the surface into crests and troughs whose long axes are perpendicular to the wind direction. The average length of jumps during saltation corresponds to the wavelength, or distance between adjacent crests, of the ripples. In ripples, the coarsest materials collect at the crests causing inverse grading. This distinguishes small ripples from dunes, where the coarsest materials are generally in the troughs. This is also a distinguishing feature between water laid ripples and aeolian ripples.

Wind-blown sand moves up the gentle upwind side of the dune by saltation or creep. Sand accumulates at the brink, the top of the slipface. When the buildup of sand at the brink exceeds the angle of repose, a small avalanche of grains slides down the slipface. Grain by grain, the dune moves downwind.

Holocene eolianite deposit on Long Island, The Bahamas. This unit is formed of wind-blown carbonate grains. (2007)

Accumulations of sediment blown by the wind into a mound or ridge, dunes have gentle upwind slopes on the windward side. The downwind portion of the dune, the lee slope, is commonly a steep avalanche slope referred to as a slipface. Dunes may have more than one slipface. The minimum height of a slipface is about 30 centimeters.

Some of the most significant experimental measurements on aeolian sand movement were performed by Ralph Alger Bagnold, a British engineer who worked in Egypt prior to World War II. Bagnold investigated the physics of particles moving through the atmosphere and deposited by wind. He recognized two basic dune types, the crescentic dune, which he called "barchan," and the linear dune, which he called longitudinal or "seif" (Arabic for "sword").

Mass Wasting

Talus cones produced by mass wasting, north shore of Isfjord, Svalbard, Norway.

Mass wasting, also known as slope movement or mass movement, is the geomorphic process by which soil, sand, regolith, and rock move downslope typically as a mass, largely under the force of gravity, but frequently affected by water and water content as in submarine environments and mudflows. Types of mass wasting include creep, slides, flows, topples, and falls, each with its own characteristic features, and taking place over timescales from seconds to years. Mass wasting occurs on both terrestrial and submarine slopes, and has been observed on Earth, Mars, Venus, and Jupiter's moon Io.

Mass wasting at Palo Duro Canyon, West Texas (2002)

When the gravitational force acting on a slope exceeds its resisting force, slope failure (mass wasting) occurs. The slope material's strength and cohesion and the amount of internal friction between material help maintain the slope's stability and are known collectively as the slope's shear strength. The steepest angle that a cohesionless slope can maintain without losing its stability is known as its angle of repose. When a slope made of loose material possesses this angle, its shear strength perfectly counterbalances the force of gravity acting upon it.

A rockfall in Grand Canyon National Park

Mass wasting may occur at a very slow rate, particularly in areas that are very dry or those areas that receive sufficient rainfall such that vegetation has stabilized the surface. It may also occur at very high speed, such as in rockslides or landslides, with disastrous consequences, both immediate and delayed, e.g., resulting from the formation of landslide dams.

Factors that change the potential of mass wasting include: change in slope angle, weakening of material by weathering, increased water content; changes in vegetation cover, and overloading.

Volcano flanks can become over-steep resulting in instability and mass wasting. It is now a recognised feature in the growth of all active volcanoes. It is seen on submarine as well as sub-aerial volcanoes - Loihi in the Hawaiian volcanic chain, Kick 'em Jenny in the Caribbean volcanic arc are two submarine volcanoes that are known to undergo mass wasting. The failure of the northern flank of Mount St Helens in 1980 showed how rapidly the flank was able to deform and fail.

The Importance of Water in Mass Wasting

Water can increase or decrease the stability of a slope depending on the amount present. Small amounts of water can strengthen soils because the surface tension of water increases soil cohesion. This allows the soil to resist erosion better than if it were dry. If too much water is present the water may act to increase the pore pressure, reducing friction, and accelerating the erosion process and resulting in different types of mass wasting (i.e. mudflows, landslides, etc.). A good example of this is to think of a sand castle. Water must be mixed with sand in order for the castle to keep its shape. If too much water is added the sand washes away, if not enough water is added the sand falls and cannot keep its shape. Water also increases the mass of the soil, this is important because an increase in mass means that there will be an increase in velocity if mass wasting is triggered.

Types of Mass Movement

Types of mass movement are distinguished based on how the soil, regolith or rock moves downslope as a whole.

Creeps

Soil creep is a long term process. The combination of small movements of soil or rock in different directions over time are directed by gravity gradually downslope. The steeper the slope, the faster the creep. The creep makes trees and shrubs curve to maintain their perpendicularity, and they can trigger landslides if they lose their root footing. The surface soil can migrate under the influence of cycles of freezing and thawing, or hot and cold temperatures, inching its way towards the bottom of the slope forming terracettes. This happens at a rate that is not noticeable to the naked eye.

Landslides

A landslide, also called a landslip, is a rapid movement of a large mass of earth and rocks down a hill or a mountainside. Little or no flowage of the materials occurs on a given slope until heavy rain and resultant lubrication by the same rainwater facilitate the movement of the materials, causing a landslide to occur. The common forms of landslides are slump, debris slide, rock slide, rock fall, debris fall and avalanche.

Flows

Movement of soil and regolith that more resembles fluid behavior is called a flow. These include avalanches, mudflows, debris flows, earth flow, lahars and sturzstroms. Water, air and ice are often involved in enabling fluidlike motion of the material.

Slump

A slipping of coherent rock material along the curved surface of a decline. Slumps involve a mass of soil or other material sliding along a curved surface (shaped like a spoon). It forms a small, crescent-shaped cliff, or abrupt scarp at the top end of the slope. There can be more than one scarp down the slope.

Falls

A fall, including rockfall, occurs where regolith cascades down a slope, but is not of sufficient volume or viscosity to behave as a flow. Falls are promoted in rocks which are characterised by presence of vertical cracks. Falls can result from undercutting by running water as well as from undercutting by waves. They usually occur at very steep slopes such as a cliff face. The rock material may be loosened by earthquakes, rain, plant-root wedging, and expanding ice, among other things. The accumulation of rock material that has fallen and resides at the base of the structure is known as talus.

Triggers of Mass Wasting

Soil and regolith remain on a hillslope only while the gravitational forces are unable to overcome the frictional forces keeping the material in place. Factors that reduce the frictional resistance relative to the downslope forces, and thus initiate slope movement, can include:

- earthquakes
- increased overburden from structures
- increased soil moisture
- reduction of roots holding the soil to bedrock
- undercutting of the slope by excavation or erosion
- weathering by frost heave
- bioturbation
- volcanic edifice over-steepening

Reducing The Impacts of Mass Wasting

- Afforestation
- Re-Afforestation
- Terracing steps on slopes (reduce angle of slope)

Soil Retrogression and Degradation

Soil retrogression and degradation are two regressive evolution processes associated with the loss of equilibrium of a stable soil. Retrogression is primarily due to soil erosion and corresponds to a phenomenon where succession reverts the land to its natural physical state. Degradation is an evolution, different from natural evolution, related to the local climate and vegetation. It is due to the replacement of primary plant communities (known as climax vegetation) by the secondary communities. This replacement modifies the humus composition and amount, and affects the formation of the soil. It is directly related to human activity. Soil degradation may also be viewed as any change or ecological disturbance to the soil perceived to be deleterious or undesirable.

General

At the beginning of soil formation, the bare rock out crops is gradually colonized by pioneer species (lichens and mosses). They are succeeded by herbaceous vegetation, shrubs and finally forest. In parallel, the first humus-bearing horizon is formed (the A horizon), followed by some mineral horizons (B horizons). Each successive stage is characterized by a certain association of soil/vegetation and environment, which defines an ecosystem.

After a certain time of parallel evolution between the ground and the vegetation, a state of steady balance is reached. This stage of development is called climax by some ecologists and "natural potential" by others. Succession is the evolution towards climax. Regardless of its name, the equilibrium stage of primary succession is the highest natural form of development that the environmental factors are capable of producing.

The cycles of evolution of soils have very variable durations, between tens, hundreds, or thousands of years for quickly evolving soils (A horizon only) to more than a million years for slowly developing soils. The same soil may achieve several successive steady state conditions during its existence, as exhibited by the Pygmy forest sequence in Mendocino County, California. Soils naturally reach a state of high productivity, from which they naturally degrade as mineral nutrients are removed from the soil system. Thus older soils are more vulnerable to the effects of induced retrogression and degradation.

Intensive tillage result on soil degradation.

Willow hedge strengthened with fascines for the limitation of runoff, north of France.

Ecological Factors Influencing Soil Formation

There are two types of ecological factors influencing the evolution of a soil (through alteration and humification). These two factors are extremely significant to explain the evolution of soils of short development.

- A first type of factor is the average climate of an area and the vegetation which is associated (biome).
- A second type of factor is more local, and is related to the original rock and local drainage. This type of factor explains appearance of specialized associations (ex peat bogs).

Biorhexistasy Theory

The destruction of the vegetation implies the destruction of evoluted soils, or a regressive evolution. Cycles of succession-regression of soils follow one another within short intervals of time (human actions) or long intervals of time (climate variations).

The climate role in the deterioration of the rocks and the formation of soils lead to the formulation of the theory of the biorhexistasy.

- In wet climate, the conditions are favorable to the deterioration of the rocks (mostly chemically), the development of the vegetation and the formation of soils; this period favorable to life is called biostasy.
- In dry climate, the rocks exposed are mostly subjected to mechanical disintegration which produces coarse detrital materials: this is referred to as rhexistasy.

Perturbations of The Balance of A Soil

When the state of balance, characterized by the ecosystem climax is reached, it tends to be maintained stable in the course of time. The vegetation installed on the ground provides the humus and ensures the ascending circulation of the matters. It protects the ground from erosion by playing the role of barrier (for example, protection from water and wind). Plants can also reduce erosion by binding the particles of the ground to their roots.

A disturbance of climax will cause retrogression, but often, secondary succession will start to guide the evolution of the system after that disturbance. Secondary succession is much faster than primary because the soil is already formed, although deteriorated and needing restoration as well.

However, when a significant destruction of the vegetation takes place (of natural origin such as an avalanche or human origin), the disturbance undergone by the ecosystem is too important. In this latter case, erosion is responsible for the destruction of the upper horizons of the ground, and is at the origin of a phenomenon of reversion to pioneer conditions. The phenomenon is called retrogression and can be partial or total (in this case, nothing remains beside bare rock). For example, the clearing of an inclined ground, subjected to violent rains, can lead to the complete destruction of the soil. Man can deeply modify the evolution of the soils by direct and brutal action, such as clearing, abusive cuts, forest pasture, litters raking. The climax vegetation is gradually replaced and the soil modified (example: replacement of leafy tree forests by moors or pines plantations). Retrogression is often related to very old human practices.

Influence of Human Activity

Soil erosion is the main factor for soil degradation and is due to several mechanisms: water erosion, wind erosion, chemical degradation and physical degradation.

Erosion is strongly related to human activity. For example, roads which increase impermeable surfaces lead to streaming and ground loss. Agriculture also accelerates soil erosion (increase of field size, correlated to hedges and ditches removal). Meadows are in regression to the profit of plowed lands. Spring cultures (sunflower, corn, beet) surfaces are increasing and leave the ground naked in winter. Sloping grounds are gradually colonized by vine. Lastly, use of herbicides leaves the ground naked between each crop. New cultural practices, such as mechanization also increases the risks of erosion. Fertilization by mineral manures rather than organic manure gradually destructure the soil. Many scientistsobserved a gradual decrease of soil organic matter content in soils, as well as a decrease of soil biological activity (in particular, in relation to chemical uses). Lastly, deforestation, in particular, is responsible for degradation of forest soils.

Agriculture increases the risk of erosion through its disturbance of vegetation by way of:

- Overgrazing of animals
- Monoculture planting

- Row cropping
- Tilling or plowing
- Crop removal
- Land-use conversion

Consequences of Soil Regression and Degradation

- yields impact: Recent increases in the human population have placed a great strain on the world's soil systems. More than 6 billion people are now using about 38% of the land area of the Earth to raise crops and livestock. Many soils suffer from various types of degradation, that can ultimately reduce their ability to produce food resources. Slight degradation refers to land where yield potential has been reduced by 10%, moderate degradation refers to a yield decrease from 10-50%. Severely degraded soils have lost more than 50% of their potential. Most severely degraded soils are located in developing countries.
- natural disasters: natural disasters such as mud flows, floods are responsible for the death of many living beings each year.
- deterioration of the water quality: the increase in the turbidity of water and the contribution of nitrogen and of phosphorus can result in eutrophication. Soils particles in surface waters are also accompanied by agricultural inputs and by some pollutants of industrial, urban and road origin (such as heavy metals). The ecological impact of agricultural inputs (such as weed killer) is known but difficult to evaluate because of the multiplicity of the products and their broad spectrum of action.
- biological diversity: soil degradation may involve perturbation of microbial communities, disappearance of the climax vegetation and decrease in animal habitat, thus leading to a biodiversity loss and animal extinction.
- economic loss: the estimated costs for land degradation are US $40 billion per year. This figure does not take into account negative externalities such as the increased use of fertilizers, loss of biodiversity and loss of unique landscapes.

Soil Enhancement, Rebuilding, and Regeneration

Problems of soil erosion can be fought, and certain practices can lead to soil enhancement and rebuilding. Even though simple, methods for reducing erosion are often not chosen because these practices outweigh the short-term benefits. Rebuilding is especially possible through the improvement of soil structure, addition of organic matter and limitation of runoff. However, these techniques will never totally succeed to restore a soil (and the fauna and flora associated to it) that took more than 1000 years to build up. Soil regeneration is the reformation of degraded soil through biological, chemical, and or physical processes.

When productivity declined in the low-clay soils of northern Thailand, farmers initially responded by adding organic matter from termite mounds, but this was unsustainable in the long-term. Scientists experimented with adding bentonite, one of the smectite family of clays, to the soil. In field trials, conducted by scientists from the International Water Management Institute in cooperation with Khon Kaen University and local farmers, this had the effect of helping retain water and nutrients. Supplementing the farmer's usual practice with a single application of 200 kg bentonite per rai (6.26 rai = 1 hectare) resulted in an average yield increase of 73%. More work showed that applying bentonite to degraded sandy soils reduced the risk of crop failure during drought years.

In 2008, three years after the initial trials, IWMI scientists conducted a survey among 250 farmers in northeast Thailand, half who had applying bentonite to their fields and half who had not. The average output for those using the clay addition was 18% higher than for non-clay users. Using the clay had enabled some farmers to switch to growing vegetables, which need more fertile soil. This helped to increase their income. The researchers estimated that 200 farmers in northeast Thailand and 400 in Cambodia had adopted the use of clays, and that a further 20,000 farmers were introduced to the new technique.

References

- Coja Isabelle and Renard, Maurice (2002) Sedimentology (translation of Sédimentologie from French) Lisse, Exton, Pennsylvania, pages 143–144, ISBN 978-90-5809-265-6
- Monroe, Wicander (2005). The Changing Earth: Exploring Geology and Evolution. Thomson Brooks/Cole. ISBN 0-495-01020-0.
- George, Rebecca; Joy, Varsha; S, Aiswarya; Jacob, Priya A. "Treatment Methods for Contaminated Soils - Translating Science into Practice" (PDF). International Journal of Education and Applied Research. Retrieved February 19, 2016.
- Retta, A.; Wagner, L.E.; Tatarko, J. (2014). "Military Vehicle Trafficking Impacts on Vegetation and Soil Bulk Density at Fort Benning, Georgia" (PDF). Transactions of the ASABE. 57 (4): 1043–1055. doi:10.13031/trans.57.10327. ISSN 2151-0032. Retrieved January 14, 2016.
- "African dust caused red soil in southern Europe". Spanish Foundation for Science and Technology. November 11, 2010. Retrieved November 2, 2012.

6

Fundamental Techniques and Approaches of Soil Management

Techniques are important components of any field of study. The following chapter elucidates the various fundamental techniques and approaches of soil management. It serves as a source to understand soil conservation, liming, soil conditioner, cover crop and nutrient management. This chapter discusses the methods of soil management in a critical manner providing key analysis to the subject matter.

Soil Conservation

Soil conservation is the prevention of soil loss from erosion or reduced fertility caused by over usage, acidification, salinization or other chemical soil contamination.

Slash-and-burn and other unsustainable methods of subsistence farming are practiced in some lesser developed areas. A sequel to the deforestation is typically large scale erosion, loss of soil nutrients and sometimes total desertification.

Erosion barriers on disturbed slope, Marin County, California

Techniques for improved soil conservation include crop rotation, cover crops, conservation tillage and planted windbreaks and affect both erosion and fertility. When plants, especially trees, die, they decay and become part of the soil. Code 330 defines standard methods recommended by the US Natural Resources Conservation Service.

Farmers have practiced soil conservation for millennia. Conservation practices fall in multiple categories:

Contour plowing, Pennsylvania 1938. The rows formed slow water run-off during rainstorms to prevent soil erosion and allows the water time to infiltrate into the soil.

Contour Plowing

Contour plowing orients crop furrows following the contour lines of the farmed area. Furrows move left and right to maintain a constant altitude, which reduces runoff. Contour plowing was practiced by the ancient Phoenicians, and is effective for slopes between two and ten percent. Contour plowing can increase crop yields from 10 to 50 percent, partially as a result from greater soil retention.

Terracing or Terrace Farming

Terracing is the practice of creating nearly level areas in a hillside area. The terraces form a series of steps, each at a higher level than the previous. Terraces are protected from erosion by other soil barriers.

Terraced farming is more common on small farms and in underdeveloped countries, since mechanized equipment is difficult to deploy in this setting. It protects the soil from its erosion. It is one of the way by which soil erosion can be stopped.

Keyline Design

Keyline design is an enhancement of contour farming, where the total watershed properties are taken into account in forming the contour lines.

Perimeter Runoff Control

Tree, shrubs and ground-cover are effective perimeter treatment for soil erosion prevention, by impeding surface flows. A special form of this perimeter or inter-row treatment is the use of a "grass way" that both channels and dissipates runoff through surface friction, impeding surface runoff and encouraging infiltration of the slowed surface water.

Windbreaks

Windbreaks are sufficiently dense rows of trees at the windward exposure of an agricultural field subject to wind erosion. Evergreen species provide year-round protection; however, as long as foliage is present in the seasons of bare soil surfaces, the effect of deciduous trees may be adequate.

Cover Crops/Crop Rotation

Cover crops such as legumes, white turnip, radishes and other species are rotated with cash crops to blanket the soil year-round and act as green manure that replenishes nitrogen and other critical nutrients. Cover crops also help suppress weeds.

Soil-Conservation Farming

Soil-conservation farming involves no-till farming, "green manures" and other soil-enhancing practices. Such farming methods attempt to mimic the biology of virgin land. They can revive damaged soil, minimize erosion, encourage plant growth, eliminate the use of nitrogen fertilizer or fungicide, produce above-average yields and protect crops during droughts or flooding. The result is less labor and lower costs that increase farmers' profits. No-till farming and cover crops act as sinks for nitrogen and other nutrients. This increases the amount of soil organic matter.

Repeated plowing/tilling degrades soil, killing its beneficial fungi and earthworms. Once damaged, soil may take multiple seasons to fully recover, even in optimal circumstances.

Critics argue that no-till and related methods are impractical and too expensive for many growers, partly because it requires new equipment. They cite advantages for conventional tilling depending on the geography, crops and soil conditions. Some farmers claimed that no-till complicates weed control, delays planting and that post-harvest residues, especially for corn, are hard to manage.

Salinity Management

Salt deposits on the former bed of the Aral Sea

Salinity in soil is caused by irrigating with salty water. Water then evaporates from the soil leaving the salt behind. Salt breaks down the soil structure, causing infertility and reduced growth.

The ions responsible for salination are: sodium (Na^{+}), potassium (K^{+}), calcium (Ca^{2+}), magnesium (Mg^{2+}) and chlorine (Cl^{-}). Salinity is estimated to affect about one third of the earth's arable land. Soil salinity adversely affects crop metabolism and erosion usually follows.

Salinity occurs on drylands from overirrigation and in areas with shallow saline water tables. Over-irrigation deposits salts in upper soil layers as a byproduct of soil infiltration; irrigation merely increases the rate of salt deposition. The best-known case of shallow saline water table capillary action occurred in Egypt after the 1970 construction of the Aswan Dam. The change in the groundwater level led to high salt concentrations in the water table. The continuous high level of the water table led to soil salination.

Use of humic acids may prevent excess salination, especially given excessive irrigation. Humic acids can fix both anions and cations and eliminate them from root zones. Planting species that can tolerate saline conditions to produce surface cover can preserve soil salinity can be reduced. Salt-tolerant plants include saltbush, a plant found in much of North America and in the Mediterranean regions of Europe.

Soil Organisms

When worms excrete egesta in the form of casts, a balanced selection of minerals and plant nutrients is made into a form accessible for root uptake. Earthworm casts are five times richer in available nitrogen, seven times richer in available phosphates and eleven times richer in available potash than the surrounding upper 150 millimetres (5.9 in) of soil. The weight of casts produced may be greater than 4.5 kg per worm per year. By burrowing, the earthworm improves soil porosity, creating channels that enhance the processes of aeration and drainage.

Yellow fungus, a mushroom that assists in organic decay.

Other important soil organisms include nematodes, mycorrizha and bacteria.

Degraded soil requires synthetic fertilizer to produce high yields. Lacking structure increases erosion and carries nitrogen and other pollutants into rivers and streams.

Each one percent increase in soil organic matter helps soil hold 20,000 gallons more water per acre.

Mineralization

To allow plants full realization of their phytonutrient potential, active mineralization of the soil is sometimes undertaken. This can involve adding crushed rock or chemical soil supplements. In either case the purpose is to combat mineral depletion. A broad range of minerals can be used, including common substances such as phosphorus and more exotic substances such as zinc and selenium. Extensive research examines the phase transitions of minerals in soil with aqueous contact.

Flooding can bring significant sediment to an alluvial plain. While this effect may not be desirable if floods endanger life or if the sediment originates from productive land, this process of addition to a floodplain is a natural process that can rejuvenate soil chemistry through mineralization and macronutrient addition.

US

For some crops, no-tillage acreage nearly doubled from 2000 to 2015. For soybeans, for example, it rose to 30 million acres in 2012 from 16.5 million acres in 1996. Use of cover crops was rising in acreage about 30 percent a year, according to surveys, though the total remains small.

Liming (Soil)

Liming is the application of calcium- and magnesium-rich materials to soil in various forms, including marl, chalk, limestone, or hydrated lime. This neutralises soil acidity and increases activity of soil bacteria. However, oversupply may result in harm to plant life. Lime is a basic chemical, the effect of it makes the soil more basic thus making acidic soils neutral.

The degree to which a given amount of lime per unit of soil volume will increase soil pH depends on the cation exchange capacity (CEC). Soils with low CEC will show a more marked pH increase than soils with high CEC. But the low-CEC soils will witness more rapid leaching of the added bases, and so will see a quicker return to original acidity unless additional liming is done.

Over-liming is most likely to occur on soil which has low CEC, such as sand which is deficient in buffering agents such as organic matter and clay.

Liming of a field in Devon

Most acid soils are saturated with aluminium rather than hydrogen ions. The acidity of the soil is therefore a result of hydrolysis of aluminium. This concept of "corrected lime potential" to define the degree of base saturation in soils became the basis for procedures now used in soil testing laboratories to determine the "lime requirement" of soils.

An agricultural study at the Faculty of Forestry in Freising, Germany that compared tree stocks 2 and 20 years after liming found that liming promotes nitrate leaching and decreases the phosphorus content of some leaves.

Soil Conditioner

A soil conditioner is a product which is added to soil to improve the soil's physical qualities, especially its ability to provide nutrition for plants. In general usage, the term "soil conditioner" is often thought of as a subset of the category soil amendments, which more often is understood to include a wide range of fertilizers and non-organic materials.

Soil conditioners can be used to improve poor soils, or to rebuild soils which have been damaged by improper management. They can make poor soils more usable, and can be used to maintain soils in peak condition.

Composition

A wide variety of materials have been described as soil conditioners due to their ability to improve soil quality. Some examples include biochar, bone meal, blood meal, coffee grounds, compost, compost tea, coir, manure, straw, peat, sphagnum moss, vermiculite, sulfur, lime, hydroabsorbant polymers, and biosolids.

Many soil conditioners come in the form of certified organic products, for people concerned with maintaining organic crops or organic gardens. Soil conditioners of almost every description are readily available from online stores or local nurseries as well as garden supply stores.

Purpose

Soil Structure

The most common use of soil conditioners is to improve soil structure. Soils tend to become compacted over time. Soil compaction impedes root growth, decreasing the ability of plants to take up nutrients and water. Soil conditioners can add more loft and texture to keep the soil loose.

Soil Nutrients

For centuries people have been adding things to poor soils to improve their ability to support healthy plant growth. Some of these materials, such as compost, clay and peat, are still used extensively today. Many soil amendments also add nutrients such as carbon and nitrogen, as well as beneficial bacteria.

Additional nutrients, such as calcium, magnesium and phosphorus, may be augmented by amendments as well. This enriches the soil, allowing plants to grow bigger and stronger.

Cation Exchange

Soil amendments can also greatly increase the cation exchange capacity of soils. Soils act as the storehouses of plant nutrients. The relative ability of soils to store one particular group of nutrients, the cations, is referred to as cation exchange capacity or CEC. The most common soil cations are calcium, magnesium, potassium, ammonium, hydrogen, and sodium.

The total number of cations a soil can hold, its total negative charge, is the soil's cation exchange capacity. The higher the CEC, the higher the negative charge and the more cations that can be held and exchanged with plant roots, providing them with the nutrition they require.

Water Retention

Soil conditioners may be used to improve water retention in dry, coarse soils which are not holding water well. The addition of organic material for instance can greatly improve the water retention abilities of sandy soils and they can be added to adjust the pH of the soil to meet the needs of specific plants or to make highly acidic or alkaline soils more usable. The possibility of using other materials to assume the role of composts and clays in improving the soil was investigated on a scientific basis earlier in the 20th century, and the term soil conditioning was coined. The criteria by which such materials are judged most often remains their cost-effectiveness, their ability to increase soil moisture for longer periods, stimulate microbiological activity, increase nutrient levels and improve plant survival rates.

The first synthetic soil conditioners were introduced in the 1950s, when the chemical hydrolysed polyacrylonitrile was the most used. Because of their ability to absorb several hundred times their own weight in water, polyacrylamides and polymethacrylates (also known as hydroabsorbent polymers, superabsorbent polymers or hydrogels) were tested in agriculture, horticulture and landscaping beginning in the 1960s.

Interest disappeared when experiments proved them to be phytotoxic due to their high acrylamide monomer residue. Although manufacturing advances later brought the monomer concentration down below the toxic level, scientific literature shows few successes in utilizing these polymers for increasing plant quality or survival. The appearance of a new generation of potentially effective tools in the early 1980s, including hydroabsorbent polymers and copolymers from the propenamide and propenamide-propenoate families, opened new perspectives.

Application

Soil conditioners may be applied in a number of ways. Some are worked into the soil with a tiller before planting. Others are applied after planting, or periodically during the growing season. Soil testing should be performed prior to applying a soil conditioner to learn more about the composition and structure of the soil. This testing will determine which conditioners will be more appropriate for the conditions.

Ecological Concerns

While adding a soil conditioner to crops or a garden can seem like a great way to get healthier plants, over-application of some amendments can cause ecological problems. For example, salts, nitrogen, metals and other nutrients that are present in many soil amendments are not productive when added in excess, and can actually be detrimental to plant health. Runoff of excess nutrients into waterways also occurs, which is harmful to the water quality and through it, the environment.

Cover Crop

A cover crop is a crop planted primarily to manage soil erosion, soil fertility, soil quality, water, weeds, pests, diseases, biodiversity and wildlife in an *agroecosystem* (Lu *et al.* 2000), an ecological system managed and largely shaped by humans across a range of intensities to produce food, feed, or fiber. Currently, not many countries are known for using the cover crop method.

Agriculture

Cover crops are of interest in sustainable agriculture as many of them improve the sustainability of agroecosystem attributes and may also indirectly improve qualities of neighboring natural ecosystems. Farmers choose to grow and manage specific cover crop types based on their own needs and goals, influenced by the biological, environmental, social, cultural, and economic factors of the food system in which they operate (Snapp *et al.* 2005). The farming practice of cover crops has been recognized as climate-smart agriculture by the White House.

Soil Erosion

Although cover crops can perform multiple functions in an agroecosystem simultaneously, they are often grown for the sole purpose of preventing soil erosion. Soil erosion is a process that can irreparably reduce the productive capacity of an agroecosystem. Dense cover crop stands physically slow down the velocity of rainfall before it contacts the soil surface, preventing soil splashing and erosive surface runoff (Romkens *et al.* 1990). Additionally, vast cover crop root networks help anchor the soil in place and increase soil porosity, creating suitable habitat networks for soil macrofauna (Tomlin *et al.* 1995).

Soil Fertility Management

One of the primary uses of cover crops is to increase soil fertility. These types of cover crops are referred to as "green manure." They are used to manage a range of soil macronutrients and micronutrients. Of the various nutrients, the impact that cover crops have on nitrogen management has received the most attention from researchers and farmers, because nitrogen is often the most limiting nutrient in crop production.

Often, green manure crops are grown for a specific period, and then plowed under before reaching full maturity in order to improve soil fertility and quality. Also the stalks left block the soil from being eroded.

Green manure crops are commonly leguminous, meaning they are part of the Fabaceae (pea) family.This family is unique in that all of the species in it set pods, such as bean, lentil, lupins and alfalfa. Leguminous cover crops are typically high in nitrogen and can often provide the required quantity of nitrogen for crop production. In conventional farming, this nitrogen is typically applied in chemical fertilizer form. This quality of cover crops is called fertilizer replacement value (Thiessen-Martens *et al.* 2005).

Another quality unique to leguminous cover crops is that they form symbiotic relationships with the rhizobial bacteria that reside in legume root nodules. Lupins is nodulated by the soil microorganism *Bradyrhizobium* sp. (Lupinus). Bradyrhizobia are encountered as microsymbionts in other leguminous crops (*Argyrolobium, Lotus, Ornithopus, Acacia, Lupinus*) of Mediterranean origin. These bacteria convert biologically unavailable atmospheric nitrogen gas (N2) to biologically available ammonium (NH+4) through the process of biological nitrogen fixation.

Prior to the advent of the Haber-Bosch process, an energy-intensive method developed to carry out industrial nitrogen fixation and create chemical nitrogen fertilizer, most nitrogen introduced to ecosystems arose through biological nitrogen fixation (Galloway *et al.* 1995). Some scientists believe that widespread biological nitrogen fixation, achieved mainly through the use of cover crops, is the only alternative to industrial nitrogen fixation in the effort to maintain or increase future food production levels (Bohlool *et al.* 1992, Peoples and Craswell 1992, Giller and Cadisch 1995). Industrial nitrogen fixation has been criticized as an unsustainable source of nitrogen for food production due to its reliance on fossil fuel energy and the environmental impacts associated with chemical nitrogen fertilizer use in agriculture (Jensen and Hauggaard-Nielsen 2003). Such widespread environmental impacts include nitrogen fertilizer losses into waterways, which can lead to eutrophication (nutrient loading) and ensuing hypoxia (oxygen depletion) of large bodies of water.

An example of this lies in the Mississippi Valley Basin, where years of fertilizer nitrogen loading into the watershed from agricultural production have resulted in a hypoxic "dead zone" off the Gulf of Mexico the size of New Jersey (Rabalais *et al.* 2002). The ecological complexity of marine life in this zone has been diminishing as a consequence (CENR 2000).

As well as bringing nitrogen into agroecosystems through biological nitrogen fixation, types of cover crops known as "catch crops" are used to retain and recycle soil nitrogen already present. The catch crops take up surplus nitrogen remaining from fertilization of the previous crop, preventing it from being lost through leaching (Morgan *et al.* 1942), or gaseous denitrification or volatilization (Thorup-Kristensen *et al.* 2003).

Catch crops are typically fast-growing annual cereal species adapted to scavenge available nitrogen efficiently from the soil (Ditsch and Alley 1991). The nitrogen tied up in catch crop biomass is released back into the soil once the catch crop is incorporated as a green manure or otherwise begins to decompose.

An example of green manure use comes from Nigeria, where the cover crop *Mucuna pruriens* (velvet bean) has been found to increase the availability of phosphorus in soil after a farmer applies rock phosphate (Vanlauwe *et al.* 2000).

Soil Quality Management

Cover crops can also improve soil quality by increasing soil organic matter levels through the input of cover crop biomass over time. Increased soil organic matter enhances soil structure, as well as the water and nutrient holding and buffering capacity of soil (Patrick *et al.* 1957). It can also lead to increased soil carbon sequestration, which has been promoted as a strategy to help offset the rise in atmospheric carbon dioxide levels (Kuo *et al.* 1997, Sainju *et al.* 2002, Lal 2003).

Soil quality is managed to produce optimum circumstances for crops to flourish. The principal factors of soil quality are soil salination, pH, microorganism balance and the prevention of soil contamination.

Water Management

By reducing soil erosion, cover crops often also reduce both the rate and quantity of water that drains off the field, which would normally pose environmental risks to waterways and ecosystems downstream (Dabney *et al.* 2001). Cover crop biomass acts as a physical barrier between rainfall and the soil surface, allowing raindrops to steadily trickle down through the soil profile. Also, as stated above, cover crop root growth results in the formation of soil pores, which in addition to enhancing soil macrofauna habitat provides pathways for water to filter through the soil profile rather than draining off the field as surface flow. With increased water infiltration, the potential for soil water storage and the recharging of aquifers can be improved (Joyce *et al.* 2002).

Just before cover crops are killed (by such practices including mowing, tilling, discing, rolling, or herbicide application) they contain a large amount of moisture. When the cover crop is incorporated into the soil, or left on the soil surface, it often increases soil moisture. In agroecosystems where water for crop production is in short supply, cover crops can be used as a mulch to conserve water by shading and cooling the soil surface. This reduces evaporation of soil moisture. In other situations farmers try to dry the soil out as quickly as possible going into the planting season. Here prolonged soil moisture conservation can be problematic.

While cover crops can help to conserve water, in temperate regions (particularly in years with below average precipitation) they can draw down soil water supply in the spring, particularly if climatic growing conditions are good. In these cases, just before crop planting, farmers often face a tradeoff between the benefits of increased cover crop growth and the drawbacks of reduced soil moisture for cash crop production that season. C/N ratio is balanced with this application.

Weed Management

Thick cover crop stands often compete well with weeds during the cover crop growth period, and can prevent most germinated weed seeds from completing their life cycle and reproducing. If the cover crop is left on the soil surface rather than incorporated into the soil as a green manure after its growth is terminated, it can form a nearly impenetrable mat. This drastically reduces light transmittance to weed seeds, which in many cases reduces weed seed germination rates (Teasdale 1993). Furthermore, even when weed seeds germinate, they often run out of stored energy for growth before building the necessary structural capacity to break through the cover crop mulch layer. This is often termed the *cover crop smother effect* (Kobayashi *et al.* 2003).

Cover crop in South Dakota

Some cover crops suppress weeds both during growth and after death (Blackshaw *et al.* 2001). During growth these cover crops compete vigorously with weeds for available space, light, and nutrients, and after death they smother the next flush of weeds by forming a mulch layer on the soil surface. For example, Blackshaw *et al.* (2001) found that when using *Melilotus officinalis* (yellow sweetclover) as a cover crop in an improved fallow system (where a fallow period is intentionally improved by any number of different management practices, including the planting of cover crops), weed biomass only constituted between 1-12% of total standing biomass at the end of the cover crop growing season. Furthermore, after cover crop termination, the yellow sweetclover residues suppressed weeds to levels 75-97% lower than in fallow (no yellow sweetclover) systems .

In addition to competition-based or physical weed suppression, certain cover crops are known to suppress weeds through allelopathy (Creamer *et al.* 1996, Singh *et al.* 2003). This occurs when certain biochemical cover crop compounds are degraded that happen to be toxic to, or inhibit seed germination of, other plant species. Some well known examples of allelopathic cover crops are *Secale cereale* (rye), *Vicia villosa* (hairy vetch), *Trifolium pratense* (red clover), *Sorghum bicolor* (sorghum-sudangrass), and species in the Brassicaceae family, particularly mustards (Haramoto and Gallandt 2004). In one study, rye cover crop residues were found to have provided between 80% and 95%

control of early season broadleaf weeds when used as a mulch during the production of different cash crops such as soybean, tobacco, corn, and sunflower (Nagabhushana *et al.* 2001).

In a recent study released by the Agricultural Research Service (ARS) scientists examined how rye seeding rates and planting patterns affected cover crop production. The results show that planting more pounds per acre of rye increased the cover crop's production as well as decreased the amount of weeds. The same was true when scientists tested seeding rates on legumes and oats; a higher density of seeds planted per acre decreased the amount of weeds and increased the yield of legume and oat production. The planting patterns, which consisted of either traditional rows or grid patterns, did not seem to make a significant impact on the cover crop's production or on the weed production in either cover crop. The ARS scientists concluded that increased seeding rates could be an effective method of weed control.

Disease Management

In the same way that allelopathic properties of cover crops can suppress weeds, they can also break disease cycles and reduce populations of bacterial and fungal diseases (Everts 2002), and parasitic nematodes (Potter *et al.* 1998, Vargas-Ayala *et al.* 2000). Species in the Brassicaceae family, such as mustards, have been widely shown to suppress fungal disease populations through the release of naturally occurring toxic chemicals during the degradation of glucosinolade compounds in their plant cell tissues (Lazzeri and Manici 2001).

Pest Management

Some cover crops are used as so-called "trap crops", to attract pests away from the crop of value and toward what the pest sees as a more favorable habitat (Shelton and Badenes-Perez 2006). Trap crop areas can be established within crops, within farms, or within landscapes. In many cases the trap crop is grown during the same season as the food crop being produced. The limited area occupied by these trap crops can be treated with a pesticide once pests are drawn to the trap in large enough numbers to reduce the pest populations. In some organic systems, farmers drive over the trap crop with a large vacuum-based implement to physically pull the pests off the plants and out of the field (Kuepper and Thomas 2002). This system has been recommended for use to help control the lygus bugs in organic strawberry production (Zalom *et al.* 2001). Another example of trap crops are nematode resistance White mustard (*Sinapis alba*) and Radish (*Raphanus sativus*). They can be grown after a main (cereal) crop and trap nematodes, for example the beet cyst nematode and Columbian root knot nematode. When grown, nematodes hatch and are attracted to the roots. After entering the roots they cannot reproduce in the root due to a hypersensitive resistance reaction of the plant. Hence the nematode population is greatly reduced, by 70-99%, depending on species and cultivation time.

Other cover crops are used to attract natural predators of pests by providing elements of their habitat. This is a form of biological control known as habitat augmentation, but achieved with the use of cover crops (Bugg and Waddington 1994). Findings on the relationship between cover crop presence and predator/pest population dynamics have been mixed, pointing toward the need for detailed information on specific cover crop types and management practices to best complement a given integrated pest management strategy. For example, the predator mite *Euseius tularensis* (Congdon) is known to help control the pest citrus thrips in Central California citrus orchards. Researchers found that the planting of several different leguminous cover crops (such as bell bean, woollypod vetch, New Zealand white clover, and Austrian winter pea) provided sufficient pollen as a feeding source to cause a seasonal increase in *E. tularensis* populations, which with good timing could potentially introduce enough predatory pressure to reduce pest populations of citrus thrips (Grafton-Cardwell *et al.* 1999).

Diversity and Wildlife

Although cover crops are normally used to serve one of the above discussed purposes, they often simultaneously improve farm habitat for wildlife. The use of cover crops adds at least one more dimension of plant diversity to a cash crop rotation. Since the cover crop is typically not a crop of value, its management is usually less intensive, providing a window of "soft" human influence on the farm. This relatively "hands-off" management, combined with the increased on-farm heterogeneity created by the establishment of cover crops, increases the likelihood that a more complex trophic structure will develop to support a higher level of wildlife diversity (Freemark and Kirk 2001).

In one study, researchers compared arthropod and songbird species composition and field use between conventionally and cover cropped cotton fields in the Southern United States. The cover cropped cotton fields were planted to clover, which was left to grow in between cotton rows throughout the early cotton growing season (stripcover cropping). During the migration and breeding season, they found that songbird densities were 7–20 times higher in the cotton fields with integrated clover cover crop than in the conventional cotton fields. Arthropod abundance and biomass was also higher in the clover cover cropped fields throughout much of the songbird breeding season, which was attributed to an increased supply of flower nectar from the clover. The clover cover crop enhanced songbird habitat by providing cover and nesting sites, and an increased food source from higher arthropod populations (Cederbaum *et al.* 2004).

Tillage

Tillage is the agricultural preparation of soil by mechanical agitation of various types, such as digging, stirring, and overturning. Examples of human-powered tilling methods using hand tools include shovelling, picking, mattock work, hoeing, and raking. Examples of draft-animal-powered or mechanized work include ploughing (overturning with moldboards or chiseling with chisel shanks), rototilling, rolling with cultipackers or other rollers, harrowing, and cultivating with cultivator shanks (teeth). Small-scale gardening and farming, for household food production or small business production, tends to use the smaller-scale methods above, whereas medium- to large-scale farming tends to use the larger-scale methods. There is a fluid continuum, however. Any type of gardening or farming, but especially larger-scale commercial types, may also use low-till or no-till methods as well.

Cultivating after an early rain

Tillage is often classified into two types, primary and secondary. There is no strict boundary between them so much as a loose distinction between tillage that is deeper and more thorough (primary) and tillage that is shallower and sometimes more selective of location (secondary). Primary tillage such as ploughing tends to produce a rough surface finish, whereas secondary tillage tends to produce a smoother surface finish, such as that required to make a good seedbed for many crops. Harrowing and rototilling often combine primary and secondary tillage into one operation.

"Tillage" can also mean the land that is tilled. The word "cultivation" has several senses that overlap substantially with those of "tillage". In a general context, both can refer to agriculture. Within agriculture, both can refer to any of the kinds of soil agitation described above. Additionally, "cultivation" or "cultivating" may refer to an even narrower sense of shallow, selective secondary tillage of row crop fields that kills weeds while sparing the crop plants.

Tillage Systems

Reduced Tillage

Reduced tillage leaves between 15 and 30% residue cover on the soil or 500 to 1000 pounds per acre (560 to 1100 kg/ha) of small grain residue during the critical erosion period. This may involve the use of a chisel plow, field cultivators, or other implements.

Plough tilling the field

Intensive Tillage

Intensive tillage leaves less than 15% crop residue cover or less than 500 pounds per acre (560 kg/ha) of small grain residue. This type of tillage is often referred to as conventional tillage but as conservational tillage is now more widely used than intensive tillage (in the United States), it is often not appropriate to refer to this type of tillage as conventional. Intensive tillage often involves multiple operations with implements such as a mold board, disk, and/or chisel plow. Then a finisher with a harrow, rolling basket, and cutter can be used to prepare the seed bed. There are many variations.

Conservation Tillage

Conservation tillage leaves at least 30% of crop residue on the soil surface, or at least 1,000 lb/ac (1,100 kg/ha) of small grain residue on the surface during the critical soil erosion period. This slows water movement, which reduces the amount of soil erosion. Conservation tillage also benefits farmers by reducing fuel consumption and soil compaction. By reducing the number of times the farmer travels over the field, farmers realize significant savings in fuel and labor. In most years since 1997, conservation tillage was used in US cropland more than intensive or reduced tillage.

However, conservation tillage delays warming of the soil due to the reduction of dark earth exposure to the warmth of the spring sun, thus delaying the planting of the next year's spring crop of corn.

- No-till - Never use a plow, disk, etc. ever again. Aims for 100% ground cover.
- Strip-Till - Narrow strips are tilled where seeds will be planted, leaving the soil in between the rows untilled.
- Mulch-till
- Rotational Tillage - Tilling the soil every two years or less often (every other year, or every third year, etc.).

- Ridge-Till
- Zone tillage - This form of conservation tillage is further explained below.

Zone Tillage

Zone tillage is a form of modified deep tillage in which only narrow strips are tilled, leaving soil in between the rows untilled. This type of tillage agitates the soil to help reduce soil compaction problems and to improve internal soil drainage.

Purpose

Zone tillage is designed to only disrupt the soil in a narrow strip directly below the crop row. In comparison to no-till, which relies on the previous year's plant residue to protect the soil and aides in postponement of the warming of the soil and crop growth in Northern climates, zone tillage creates approximately a 5-inch-wide strip that simultaneously breaks up plow pans, assists in warming the soil and helps to prepare a seedbed. When combined with cover crops, zone tillage helps replace lost organic matter, slows the deterioration of the soil, improves soil drainage, increases soil water and nutrient holding capacity, and allows necessary soil organisms to survive.

Usage

It has been successfully used on farms in the mid-west and west for over 40 years and is currently used on more than 36% of the U.S. farmland. Some specific states where zone tillage is currently in practice are Pennsylvania, Connecticut, Minnesota, Indiana, Wisconsin, and Illinois.

Unfortunately, there aren't consistent yield results in the Northern Cornbelt states; however, there is still interest in deep tillage within the agriculture industry. In areas that are not well-drained, deep tillage may be used as an alternative to installing more expensive tile drainage.

Effects of Tillage

Positive

Plowing:

- Loosens and aerates the top layer of soil, which facilitates planting the crop
- Helps mix harvest residue, organic matter (humus), and nutrients evenly into the soil
- Mechanically destroys weeds
- Dries the soil before seeding (in wetter climates tillage aids in keeping the soil drier)
- When done in autumn, helps exposed soil crumble over winter through frosting

and defrosting, which helps prepare a smooth surface for spring planting

Negative

- Dries the soil before seeding
- Soil loses a lot of nutrients, like nitrogen and fertilizer, and its ability to store water
- Decreases the water infiltration rate of soil. (Results in more runoff and erosion since the soil absorbs water slower than before)
- Tilling the soil results in dislodging the cohesiveness of the soil particles thereby inducing erosion.
- Chemical runoff
- Reduces organic matter in the soil
- Reduces microbes, earthworms, ants, etc.
- Destroys soil aggregates
- Compaction of the soil, also known as a tillage pan
- Eutrophication (nutrient runoff into a body of water)
- Can attract slugs, cut worms, army worms, and harmful insects to the left over residues.
- Crop diseases can be harbored in surface residues

General Comments

- The type of implement makes the most difference, although other factors can have an effect.
- Tilling in absolute darkness (night tillage) might reduce the number of weeds that sprout following the tilling operation by half. Light is necessary to break the dormancy of some weed species' seed, so if fewer seeds are exposed to light during the tilling process, fewer will sprout. This may help reduce the amount of herbicides needed for weed control.
- Greater speeds, when using certain tillage implements (disks and chisel plows), lead to more intensive tillage (i.e., less residue is on the soil surface).
- Increasing the angle of disks causes residues to be buried more deeply. Increasing their concavity makes them more aggressive.
- Chisel plows can have spikes or sweeps. Spikes are more aggressive.
- Percentage residue is used to compare tillage systems because the amount of

crop residue affects the soil loss due to erosion.

Definitions

Primary tillage loosens the soil and mixes in fertilizer and/or plant material, resulting in soil with a rough texture.

Secondary tillage produces finer soil and sometimes shapes the rows, preparing the seed bed. It also provides weed control throughout the growing season during the maturation of the crop plants, unless such weed control is instead achieved with low-till or no-till methods involving herbicides.

- The seed bed preparation can be done with harrows (of which there are many types and subtypes), dibbles, hoes, shovels, rotary tillers, subsoilers, ridge- or bed-forming tillers, rollers, or cultivators.
- The weed control, to the extent that it is done via tillage, is usually achieved with cultivators or hoes, which disturb the top few centimeters of soil around the crop plants but with minimal disturbance of the crop plants themselves. The tillage kills the weeds via 2 mechanisms: uprooting them, burying their leaves (cutting off their photosynthesis), or a combination of both. Weed control both prevents the crop plants from being outcompeted by the weeds (for water and sunlight) and prevents the weeds from reaching their seed stage, thus reducing future weed population aggressiveness.

History of Tilling

Tilling was first performed via human labor, sometimes involving slaves. Hoofed animals could also be used to till soil via trampling. The wooden plow was then invented. It could be pulled by mule, ox, elephant, water buffalo, or similar sturdy animal. Horses are generally unsuitable, though breeds such as the scyne could work. The steel plow allowed farming in the American Midwest, where tough prairie grasses and rocks caused trouble. Soon after 1900, the farm tractor was introduced, which eventually made modern large-scale agriculture possible.

Tilling with Hungarian Grey cattles

Alternatives to Tilling

Modern agricultural science has greatly reduced the use of tillage. Crops can be grown for several years without any tillage through the use of herbicides to control weeds, crop varieties that tolerate packed soil, and equipment that can plant seeds or fumigate the soil without really digging it up. This practice, called no-till farming, reduces costs and environmental change by reducing soil erosion and diesel fuel usage.

Site Preparation of Forest Land

Site preparation is any of various treatments applied to a site in order to ready it for seeding or planting. The purpose is to facilitate the regeneration of that site by the chosen method. Site preparation may be designed to achieve, singly or in any combination: improved access, by reducing or rearranging slash, and amelioration of adverse forest floor, soil, vegetation, or other biotic factors. Site preparation is undertaken to ameliorate one or more constraints that would otherwise be likely to thwart the objectives of management. A valuable bibliography on the effects of soil temperature and site preparation on subalpine and boreal tree species has been prepared by McKinnon et al. (2002).

Site preparation is the work that is done before a forest area is regenerated. Some types of site preparation are burning.

Burning

Broadcast burning is commonly used to prepare clearcut sites for planting, e.g., in central British Columbia, and in the temperate region of North America generally.

Prescribed burning is carried out primarily for slash hazard reduction and to improve site conditions for regeneration; all or some of the following benefits may accrue:

- α) Reduction of logging slash, plant competition, and humus prior to direct seeding, planting, scarifying or in anticipation of natural seeding in partially cut stands or in connection with seed-tree systems.
- β) Reduction or elimination of unwanted forest cover prior to planting or seeding, or prior to preliminary scarification thereto.
- χ) Reduction of humus on cold, moist sites to favour regeneration.
- δ) Reduction or elimination of slash, grass, or brush fuels from strategic areas around forested land to reduce the chances of damage by wildfire.

Prescribed burning for preparing sites for direct seeding was tried on a few occasions in Ontario, but none of the burns was hot enough to produce a seedbed that was adequate without supplementary mechanical site preparation.

Changes in soil chemical properties associated with burning include significantly increased pH, which Macadam (1987) in the Sub-boreal Spruce Zone of central British Columbia found persisting more than a year after the burn. Average fuel consumption was 20 to 24 t/ha and the forest floor depth was reduced by 28% to 36%. The increases correlated well with the amounts of slash (both total and ≥7 cm diameter) consumed. The change in pH depends on the severity of the burn and the amount consumed; the increase can be as much as 2 units, a 100-fold change. Deficiencies of copper and iron in the foliage of white spruce on burned clearcuts in central British Columbia might be attributable to elevated pH levels.

Even a broadcast slash fire in a clearcut does not give a uniform burn over the whole area. Tarrant (1954), for instance, found only 4% of a 140-ha slash burn had burned severely, 47% had burned lightly, and 49% was unburned. Burning after windrowing obviously accentuates the subsequent heterogeneity.

Marked increases in exchangeable calcium also correlated with the amount of slash at least 7 cm in diameter consumed. Phosphorus availability also increased, both in the forest floor and in the 0 cm to 15 cm mineral soil layer, and the increase was still evident, albeit somewhat diminished, 21 months after burning. However, in another study in the same Sub-boreal Spruce Zone found that although it increased immediately after the burn, phosphorus availability had dropped to below pre-burn levels within 9 months.

Nitrogen will be lost from the site by burning, though concentrations in remaining forest floor were found by Macadam (1987) to have increased in 2 of 6 plots, the others showing decreases. Nutrient losses may be outweighed, at least in the short term, by improved soil microclimate through the reduced thickness of forest floor where low soil temperatures are a limiting factor.

The *Picea/Abies* forests of the Alberta foothills are often characterized by deep accumulations of organic matter on the soil surface and cold soil temperatures, both of which make reforestation difficult and result in a general deterioration in site productivity; Endean and Johnstone (1974) describe experiments to test prescribed burning as a means of seedbed preparation and site amelioration on representative clear-felled *Picea/Abies* areas. Results showed that, in general, prescribed burning did not reduce organic layers satisfactorily, nor did it increase soil temperature, on the sites tested. Increases in seedling establishment, survival, and growth on the burned sites were probably the result of slight reductions in the depth of the organic layer, minor increases in soil temperature, and marked improvements in the efficiency of the planting crews. Results also suggested that the process of site deterioration has not been reversed by the burning treatments applied.

Ameliorative Intervention

Slash weight (the oven-dry weight of the entire crown and that portion of the stem < 4 inches in diameter) and size distribution are major factors influencing the forest fire hazard on harvested sites. Forest managers interested in the application of prescribed burning for hazard reduction and silviculture, were shown a method for quantifying the slash load by Kiil (1968). In west-central Alberta, he felled, measured, and weighed 60 white spruce, graphed (a) slash weight per merchantable unit volume against diameter at breast height (dbh), and (b) weight of fine slash (<1.27 cm) also against dbh, and produced a table of slash weight and size distribution on one acre of a hypothetical stand of white spruce. When the diameter distribution of a stand is unknown, an estimate of slash weight and size distribution can be obtained from average stand diameter, number of trees per unit area, and merchantable cubic foot volume. The sample trees in Kiil's study had full symmetrical crowns. Densely growing trees with short and often irregular crowns would probably be overestimated; open-grown trees with long crowns would probably be underestimated.

The need to provide shade for young outplants of Engelmann spruce in the high Rocky Mountains is emphasized by the U.S. Forest Service. Acceptable planting spots are defined as microsites on the north and east sides of down logs, stumps, or slash, and lying in the shadow cast by such material. Where the objectives of management specify more uniform spacing, or higher densities, than obtainable from an existing distribution of shade-providing material, redistribution or importing of such material has been undertaken.

Access

Site preparation on some sites might be done simply to facilitate access by planters, or to improve access and increase the number or distribution of microsites suitable for planting or seeding.

Wang et al. (2000) determined field performance of white and black spruces 8 and 9 years after outplanting on boreal mixedwood sites following site preparation (Donaren disc trenching versus no trenching) in 2 plantation types (open versus sheltered) in southeastern Manitoba. Donaren trenching slightly reduced the mortality of black spruce but significantly increased the mortality of white spruce. Significant difference in height was found between open and sheltered plantations for black spruce but not for white spruce, and root collar diameter in sheltered plantations was significantly larger than in open plantations for black spruce but not for white spruce. Black spruce open plantation had significantly smaller volume (97 cm^3) compared with black spruce sheltered (210 cm^3), as well as white spruce open (175 cm^3) and sheltered (229 cm^3) plantations. White spruce open plantations also had smaller volume than white spruce sheltered plantations. For transplant stock, strip plantations had a significantly higher volume (329 cm^3) than open plantations (204 cm^3). Wang et al. (2000) recommended that sheltered plantation site preparation should be used.

Mechanical

Up to 1970, no "sophisticated" site preparation equipment had become operational in Ontario, but the need for more efficacious and versatile equipment was increasingly recognized. By this time, improvements were being made to equipment originally developed by field staff, and field testing of equipment from other sources was increasing.

According to J. Hall (1970), in Ontario at least, the most widely used site preparation technique was post-harvest mechanical scarification by equipment front-mounted on a bulldozer (blade, rake, V-plow, or teeth), or dragged behind a tractor (Imsett or S.F.I. scarifier, or rolling chopper). Drag type units designed and constructed by Ontario's Department of Lands and Forests used anchor chain or tractor pads separately or in combination, or were finned steel drums or barrels of various sizes and used in sets alone or combined with tractor pad or anchor chain units.

J. Hall's (1970) report on the state of site preparation in Ontario noted that blades and rakes were found to be well suited to post-cut scarification in tolerant hardwood stands for natural regeneration of yellow birch. Plows were most effective for treating dense brush prior to planting, often in conjunction with a planting machine. Scarifying teeth, e.g., Young's teeth, were sometimes used to prepare sites for planting, but their most effective use was found to be preparing sites for seeding, particularly in backlog areas carrying light brush and dense herbaceous growth. Rolling choppers found application in treating heavy brush but could be used only on stone-free soils. Finned drums were commonly used on jack pine–spruce cutovers on fresh brushy sites with a deep duff layer and heavy slash, and they needed to be teamed with a tractor pad unit to secure good distribution of the slash. The S.F.I. scarifier, after strengthening, had been "quite successful" for 2 years, promising trials were under way with the cone scarifier and barrel ring scarifier, and development had begun on a new flail scarifier for use on sites with shallow, rocky soils. Recognition of the need to become more effective and efficient in site preparation led the Ontario Department of Lands and Forests to adopt the policy of seeking and obtaining for field testing new equipment from Scandinavia and elsewhere that seemed to hold promise for Ontario conditions, primarily in the north. Thus, testing was begun of the Brackekultivator from Sweden and the Vako-Visko rotary furrower from Finland.

Mounding

Site preparation treatments that create raised planting spots have commonly improved outplant performance on sites subject to low soil temperature and excess soil moisture. Mounding can certainly have a big influence on soil temperature. Draper et al. (1985), for instance, documented this as well as the effect it had on root growth of outplants (Table 30).

The mounds warmed up quickest, and at soil depths of 0.5 cm and 10 cm averaged 10 and 7 °C higher, respectively, than in the control. On sunny days, daytime surface temperature maxima on the mound and organic mat reached 25 °C to 60 °C, depending on soil wetness and shading. Mounds reached mean soil temperatures of 10 °C at 10 cm depth 5 days after planting, but the control did not reach that temperature until 58 days after planting. During the first growing season, mounds had 3 times as many days with a mean soil temperature greater than 10 °C than did the control microsites.

Draper et al.'s (1985) mounds received 5 times the amount of photosynthetically active radiation (PAR) summed over all sampled microsites throughout the first growing season; the control treatment consistently received about 14% of daily background PAR, while mounds received over 70%. By November, fall frosts had reduced shading, eliminating the differential. Quite apart from its effect on temperature, incident radiation is also important photosynthetically. The average control microsite was exposed to levels of light above the compensation point for only 3 hours, i.e., one-quarter of the daily light period, whereas mounds received light above the compensation point for 11 hours, i.e., 86% of the same daily period. Assuming that incident light in the 100-600 $\mu Em^{-2}s^{-1}$ intensity range is the most important for photosynthesis, the mounds received over 4 times the total daily light energy that reached the control microsites.

Orientation of Linear Site Preparation, E.g., Disk-Trenching

With linear site preparation, orientation is sometimes dictated by topography or other considerations, but the orientation can often be chosen. It can make a difference. A disk-trenching experiment in the Sub-boreal Spruce Zone in interior British Columbia investigated the effect on growth of young outplants (lodgepole pine) in 13 microsite planting positions: berm, hinge, and trench in each of north, south, east, and west aspects, as well as in untreated locations between the furrows. Tenth-year stem volumes of trees on south, east, and west-facing microsites were significantly greater than those of trees on north-facing and untreated microsites. However, planting spot selection was seen to be more important overall than trench orientation.

In a Minnesota study, the N–S strips accumulated more snow but snow melted faster than on E–W strips in the first year after felling. Snow-melt was faster on strips near the centre of the strip-felled area than on border strips adjoining the intact stand. The strips, 50 feet (15.24 m) wide, alternating with uncut strips 16 feet (4.88 m) wide, were felled in a *Pinus resinosa* stand, aged 90 to 100 years.

Mulch

A mulch is a layer of material applied to the surface of an area of soil. Its purpose is any or all of the following:

- to conserve moisture
- to improve the fertility and health of the soil
- to reduce weed growth
- to enhance the visual appeal of the area

A mulch is usually but not exclusively organic in nature. It may be permanent (e.g. plastic sheeting) or temporary (e.g. bark chips). It may be applied to bare soil, or around existing plants. Mulches of manure or compost will be incorporated naturally into the soil by the activity of worms and other organisms. The process is used both in commercial crop production and in gardening, and when applied correctly can dramatically improve soil productivity.

Uses

Many materials are used as mulches, which are used to retain soil moisture, regulate soil temperature, suppress weed growth, and for aesthetics. They are applied to the soil surface, around trees, paths, flower beds, to prevent soil erosion on slopes, and in production areas for flower and vegetable crops. Mulch layers are normally two inches or more deep when applied.

They are applied at various times of the year depending on the purpose. Towards the beginning of the growing season mulches serve initially to warm the soil by helping it retain heat which is lost during the night. This allows early seeding and transplanting of certain crops, and encourages faster growth. As the season progresses, mulch stabilizes the soil temperature and moisture, and prevents the growing of weeds from seeds. In temperate climates, the effect of mulch is dependent upon the time of year they are applied and when applied in fall and winter, are used to delay the growth of perennial plants in the spring or prevent growth in winter during warm spells, which limits freeze thaw damage.

The effect of mulch upon soil moisture content is complex. Mulch forms a layer between the soil and the atmosphere which prevents sunlight from reaching the soil surface, thus reducing evaporation. However, mulch can also prevent water from reaching the soil by absorbing or blocking water from light rains.

In order to maximise the benefits of mulch, while minimizing its negative influences, it is often applied in late spring/early summer when soil temperatures have risen sufficiently, but soil moisture content is still relatively high. However, permanent mulch is also widely used and valued for its simplicity, as popularized by author Ruth Stout, who said, "My way is simply to keep a thick mulch of any vegetable matter that rots on both sides of my vegetable and flower garden all year long. As it decays and enriches the soils, I add more."

Plastic mulch used in large-scale commercial production is laid down with a tractor-drawn or standalone layer of plastic mulch. This is usually part of a sophisticated mechanical process, where raised beds are formed, plastic is rolled out on top, and seedlings are transplanted through it. Drip irrigation is often required, with drip tape laid under the plastic, as plastic mulch is impermeable to water.

Materials

Rubber mulch nuggets in a playground. The white fibers are nylon cords, which are present in the tires from which the mulch is made.

Aged Compost mulch on a flower bed

Shredded wood used as mulch. This type of mulch is often dyed to improve its appearance in the landscape.

Crushed stone mulch

Pine needles used as mulch. Also called "pinestraw" in the southern US.

Spring daffodils push through shredded wood mulch

Materials used as mulches vary and depend on a number of factors. Use takes into consideration availability, cost, appearance, the effect it has on the soil—including chemical reactions and pH, durability, combustibility, rate of decomposition, how clean it is—some can contain weed seeds or plant pathogens.

A variety of materials are used as mulch:

- Organic residues: grass clippings, leaves, hay, straw, kitchen scraps comfrey, shredded bark, whole bark nuggets, sawdust, shells, woodchips, shredded newspaper, cardboard, wool, animal manure, etc. Many of these materials also act as a direct composting system, such as the mulched clippings of a mulching lawn mower, or other organics applied as sheet composting.

- Compost: fully composted materials are used to avoid possible phytotoxicity problems. Materials that are free of seeds are ideally used, to prevent weeds being introduced by the mulch.
- Old carpet (synthetic or natural): makes a free, readily available mulch.
- Rubber mulch: made from recycled tire rubber.
- Plastic mulch: crops grow through slits or holes in thin plastic sheeting. This method is predominant in large-scale vegetable growing, with millions of acres cultivated under plastic mulch worldwide each year (disposal of plastic mulch is cited as an environmental problem).
- Rock and gravel can also be used as a mulch. In cooler climates the heat retained by rocks may extend the growing season.

In some areas of the United States, such as central Pennsylvania and northern California, mulch is often referred to as "tanbark", even by manufacturers and distributors. In these areas, the word "mulch" is used specifically to refer to very fine tanbark or peat moss.

Organic Mulches

Organic mulches decay over time and are temporary. The way a particular organic mulch decomposes and reacts to wetting by rain and dew affects its usefulness.

Some mulches such as straw, peat, sawdust and other wood products may for a while negatively affect plant growth because of their wide carbon to nitrogen ratio, because bacteria and fungi that decompose the materials remove nitrogen from the surrounding soil for growth. However, whether this effect has any practical impact on gardens is disputed by researchers and the experience of gardeners. Organic mulches can mat down, forming a barrier that blocks water and air flow between the soil and the atmosphere. Vertically applied organic mulches can wick water from the soil to the surface, which can dry out the soil. Mulch made with wood can contain or feed termites, so care must be taken about not placing mulch too close to houses or building that can be damaged by those insects. Some mulch manufacturers recommend putting mulch several inches away from buildings.

Commonly available organic mulches include:

Leaves

- *Leaves* from deciduous trees, which drop their foliage in the autumn/fall. They tend to be dry and blow around in the wind, so are often chopped or shredded before application. As they decompose they adhere to each other but also allow water and moisture to seep down to the soil surface. Thick layers of entire leaves, especially of maples and oaks, can form a soggy mat in winter and spring which can impede the new growth lawn grass and other plants. Dry leaves are used as winter mulches to protect plants from freezing and thawing in areas with cold winters, they are normally removed during spring.

Grass Clippings

- *Grass clippings*, from mowed lawns are sometimes collected and used elsewhere as mulch. Grass clippings are dense and tend to mat down, so are mixed with tree leaves or rough compost to provide aeration and to facilitate their decomposition without smelly putrefaction. Rotting fresh grass clippings can damage plants; their rotting often produces a damaging buildup of trapped heat. Grass clippings are often dried thoroughly before application, which mediates against rapid decomposition and excessive heat generation. Fresh green grass clippings are relatively high in nitrate content, and when used as a mulch, much of the nitrate is returned to the soil, but the routine removal of grass clippings from the lawn results in nitrogen deficiency for the lawn.

Peat Moss

- *Peat moss*, or sphagnum peat, is long lasting and packaged, making it convenient and popular as a mulch. When wetted and dried, it can form a dense crust that does not allow water to soak in. When dry it can also burn, producing a smoldering fire. It is sometimes mixed with pine needles to produce a mulch that is friable. It can also lower the pH of the soil surface, making it useful as a mulch under acid loving plants.

However peat bogs are a valuable wildlife habitat, and peat is also one of the largest stores of carbon (in Britain, out of a total estimated 9952 million tonnes of carbon in British vegetation and soils, 6948 million tonnes carbon are estimated to be in Scottish, mostly peatland, soils), so gardeners who wish to protect the environment will choose more sustainable alternatives.

Wood Chips

- *Wood chips* are a byproduct of the pruning of trees by arborists, utilities and parks; they are used to dispose of bulky waste. Tree branches and large stems are rather coarse after chipping and tend to be used as a mulch at least three inches thick. The chips are used to conserve soil moisture, moderate soil temperature and suppress weed growth. The decay of freshly produced chips from recently living woody plants, consumes nitrate; this is often off set with a light application of a high-nitrate fertilizer. Wood chips are most often used under trees and shrubs. When used around soft stemmed plants, an unmulched zone is left around the plant stems to prevent stem rot or other possible diseases. They are often used to mulch trails, because they are readily produced with little additional cost outside of the normal disposal cost of tree maintenance. Wood chips come in various colors.

Woodchip Mulch

- *Woodchip mulch* is a byproduct of reprocessing used (untreated) timber (usually packaging pallets), to dispose of wood waste by creating woodchip mulch. The chips are used to conserve soil moisture, moderate soil temperature and suppress weed growth. Woodchip mulch is often used under trees, shrubs or large planting areas and can last much longer than arborist mulch. In addition, many consider woodchip mulch to be visually appealing, as it comes in various colors. Woodchips can also be reprocessed into playground woodchip to be used as an impact-attenuating playground surfacing.

Bark Chips

Bark chips

- *Bark chips* of various grades are produced from the outer corky bark layer of timber trees. Sizes vary from thin shredded strands to large coarse blocks. The finer types are very attractive but have a large exposed surface area that leads to quicker decay. Layers two or three inches deep are usually used, bark is relativity inert and its decay does not demand soil nitrates. Bark chips are also available in various colors.

Straw Mulch/Field Hay/Salt Hay

Permaculture garden with a fruit tree, herbs, flowers and vegetables mulched with hay

- *Straw mulch* or *field hay* or *salt hay* are lightweight and normally sold in compressed bales. They have an unkempt look and are used in vegetable gardens and as a winter covering. They are biodegradable and neutral in pH. They have good moisture retention and weed controlling properties but also are more likely to be contaminated with weed seeds. Salt hay is less likely to have weed seeds than field hay. Straw mulch is also available in various colors.

Cardboard/Newspaper

- *Cardboard* or *newspaper* can be used as mulches. These are best used as a base layer upon which a heavier mulch such as compost is placed to prevent the lighter cardboard/newspaper layer from blowing away. By incorporating a layer of cardboard/newspaper into a mulch, the quantity of heavier mulch can be reduced, whilst improving the weed suppressant and moisture retaining properties of the mulch. However, additional labour is expended when planting through a mulch containing a cardboard/newspaper layer, as holes must be cut for each plant. Sowing seed through mulches containing a cardboard/newspaper layer is impractical. Application of newspaper mulch in windy weather can be facilitated by briefly pre-soaking the newspaper in water to increase its weight.

Carpet

- Synthetic carpet that is composed of artificial fibers may be removed after planting to prevent fibers taking a long time to decompose, whereas carpet made from natural fibers may be kept in place, blocking competition from weeds. Rain is absorbed by carpet and then slowly released into the soil, reducing watering needs.

Colored Mulch

Some organic mulches are colored red, brown, black, and other colors. Isopropanolamine, specifically 1-Amino-2-propanol or DOW™ monoisopropanolamine, may be used as a pigment dispersant and color fastener in these mulches. Types of mulch which can be dyed include: wood chips, bark chips (barkdust) and pine straw. Colored mulch is made by dyeing the mulch in a water-based solution of colorant and chemical binder. When colored mulch first entered the market, most formulas were suspected to contain toxic, heavy metals and other contaminates. Today, "current investigations indicate that mulch colorants pose no threat to people, pets or the environment. The dyes currently used by the mulch and soil industry are similar to those used in the cosmetic and other manufacturing industries (i.e., iron oxide)," as stated by the Mulch and Soil Council. Colored mulch can be applied anywhere non-colored mulch is used (such as large bedded areas or around plants) and features many of the same gardening benefits as traditional mulch, such as improving soil productivity and retaining moisture. As mulch decomposes, just as with non-colored mulch, more mulch may need to be added to continue providing benefits to the soil and plants. However, if mulch is faded, spraying dye to previously spread mulch in order to restore color is an option.

Anaerobic (Sour) Mulch

Mulch normally smells like freshly cut wood, but sometimes develops a toxicity that causes it to smell like vinegar, ammonia, sulfur or silage. This happens when material with ample nitrogen content is not rotated often enough and it forms pockets of increased decomposition. When this occurs, the process may become anaerobic and produce these phytotoxic materials in small quantities. Once exposed to the air, the process quickly reverts to an aerobic process, but these toxic materials may be present for a period of time. If the mulch is placed around plants before the toxicity has had a chance to dissipate, then the plants could very likely be damaged or killed depending on their hardiness. Plants that are predominantly low to the ground or freshly planted are the most susceptible, and the phytotoxicity may prevent germination of some seeds.

If sour mulch is applied and there is plant kill, the best thing to do is to water the mulch heavily. Water dissipates the chemicals faster and refreshes the plants. Removing the offending mulch may have little effect, because by the time plant kill is noticed, most of the toxicity is already dissipated. While testing after plant kill will not likely turn up anything, a simple pH check may reveal high acidity, in the range of 3.8 to 5.6 instead of the normal range of 6.0 to 7.2. Finally, placing a bit of the offending mulch around another plant to check for plant kill will verify if the toxicity has departed. If the new plant is also killed, then sour mulch is probably not the problem.

Groundcovers (Living Mulches)

Groundcovers are plants which grow close to the ground, under the main crop, to slow the development of weeds and provide other benefits of mulch. They are usually fast-growing plants that continue growing with the main crops. By contrast, cover crops are incorporated into the soil or killed with herbicides. However, live mulches also may need to be mechanically or chemically killed eventually to prevent competition with the main crop.

Some groundcovers can perform additional roles in the garden such as nitrogen fixation in the case of clovers, dynamic accumulation of nutrients from the subsoil in the case of creeping comfrey (*Symphytum ibericum*), and even food production in the case of *Rubus tricolor*.

On-Site Mulch Production

Owing to the great bulk of mulch which is often required on a site, it is often impractical and expensive to source and import sufficient mulch materials. An alternative to importing mulch materials is to grow them on site in a "mulch garden" - an area of the site dedicated entirely to the production of mulch which is then transferred to the growing area. Mulch gardens should be sited as close as possible to the growing area so as to facilitate transfer of mulch materials.

Mulching (Composting) Over Unwanted Plants

Sufficient mulch over plants will destroy them, and may be more advantageous than using herbicide, cutting, mowing, pulling, raking, or tilling. The higher the temperature that this "mulch" is composted, the quicker the reduction of undesirable materials. "Undesirable materials" may include living seed, plant "trash", as well as pathogens such as from animal feces, urine (e.g. hantavirus), fleas, lice, ticks, etc.

In some ways this improves the soil by attracting and feeding earthworms, and adding humus. Earthworms "till" the soil, and their feces are among the best fertilizers and soil conditioners.

Urine may be toxic to plants if applied to growing areas undiluted.

Polypropylene and Polyethylene Mulch

Polypropylene mulch is made up of polypropylene polymers where polyethylene mulch is made up of polyethylene polymers. These mulches are commonly used in many plastics. Polyethylene is used mainly for weed reduction, where polypropylene is used mainly on perennials. This mulch is placed on top of the soil and can be done by machine or hand with pegs to keep the mulch tight against the soil. This mulch can prevent soil erosion, reduce weeding, conserve soil moisture, and increase temperature of the soil. Ultimately this can reduce the amount of work a farmer may have to do, and the amount of herbicides applied during the growing period. The black and clear mulches capture sunlight and warm the soil increasing the growth rate. White and other reflective colours will also warm the soil, but they do not suppress weeds as well. This mulch may require other sources of obtaining water such as drip irrigation since it can reduce the amount of water that reaches the soil. This mulch needs to be manually removed at the end of the season since when it starts to break down it breaks down into smaller pieces. If the mulch is not removed before it starts to break down eventually it will break down into ketones and aldehydes polluting the soil. This mulch is technically biodegradable but does not break down into the same materials the more natural biodegradable mulch does.

Biodegradable Mulch

Quality biodegradable mulches are made out of plant starches and sugars or polyester fibres. These starches can come from plants such as wheat and corn. These mulch films may be a bit more permeable allowing more water into the soil. This mulch can prevent soil erosion, reduce weeding, conserve soil moisture, and increase temperature of the soil. Ultimately this can reduce the amount of herbicides used and manual labour farmers may have to do throughout the growing season. At the end of the season these mulches will start to break down from heat. Microorganisms in the soil break down the mulch into two components, water and CO2, leaving no toxic residues behind. This source of mulch is even less manual labour since it does not need to be removed at the end of the season and can actually be tilled into the soil. With this mulch its important to take into consideration that its mulch is more delicate then other kinds. It should be placed on a day which is not too hot and with less tension then other synthetic mulches. These also can be placed by machine or hand but its ideal to have a more starchy mulch that will allow it to stick to the soil better.

Nutrient Management

Nutrient management is the science and art directed to link soil, crop, weather, and hydrologic factors with cultural, irrigation, and soil and water conservation practices to achieve the goals of optimizing nutrient use efficiency, yields, crop quality, and economic returns, while reducing off-site transport of nutrients that may impact the environment. Nutrient management is the skillful task of matching a specific field soil, climate, and crop management conditions to rate, source, timing, and place (commonly known as the 4R nutrient stewardship) of nutrient application.

Nitrogen fertilizer being applied to growing corn (maize) in a contoured, no-tilled field in Iowa.

Some important factors that need to be considered when managing nutrients include (a) the application of nutrients considering the achievable optimum yields and, in some cases, crop quality; (b) the management, application, and timing of nutrients using a budget based on all sources and sinks active at the site; and (c) the management of soil, water, and crop to minimize the off-site transport of nutrients from nutrient leaching out of the root zone, surface runoff, and volatilization (or other gas exchanges).

There can be potential interactions because of differences in nutrient pathways and dynamics. For instance, practices that reduce the off-site surface transport of a given nutrient may increase the leaching losses of other nutrients. These complex dynamics present nutrient managers the difficult task of integrating soil, crop, weather, hydrology, and management practices to achieve the best balance for maximizing profit while contributing to the conservation of our biosphere.

Nutrient Management Plan

A crop nutrient management plan is a tool that farmers can use to increase the efficiency of all the nutrient sources a crop uses while reducing production and environmental risk, ultimately increasing profit. It is generally agreed that there are ten fundamental components of a Crop Nutrient Management Plan. Each component is critical to helping analyze each field and improve nutrient efficiency for the crops grown. These components include:

Manure spreader

- *Field Map:* The map, including general reference points (such as streams, residences, wellheads etc.), number of acres, and soil types is the base for the rest of the plan.
- *Soil Test:* How much of each nutrient (N-P-K and other critical elements such as pH and organic matter) is in the soil profile? The soil test is a key component needed for developing the nutrient rate recommendation.
- *Crop Sequence:* Did the crop that grew in the field last year (and in many cases two or more years ago) fix nitrogen for use in the following years? Has long-term no-till increased organic matter? Did the end-of-season stalk test show a nutrient deficiency? These factors also need to be factored into your plan.
- *Estimated Yield:* Factors that affect yield are numerous and complex. A field's soils, drainage, insect, weed and disease pressure, rotation and many other factors differentiate one field from another. This is why using historic yields is important in developing yield estimates for next year. Accurate yield estimates can dramatically improve nutrient use efficiency.
- *Sources and Forms:* The sources and forms of available nutrients can vary from farm-to-farm and even field-to-field. For instance, manure fertility analysis, storage practices and other factors will need to be included in a nutrient

management plan. Manure nutrient tests/analysis are one way to determine the fertility of it. Nitrogen fixed from a previous year's legume crop and residual affects of manure also effects rate recommendations. Many other nutrient sources should also be factored into this plan.

- *Sensitive Areas:* What's out of the ordinary about a field's plan? Is it irrigated? Next to a stream or lake? Especially sandy in one area? Steep slope or low area? Manure applied in one area for generations due to proximity of dairy barn? Extremely productive—or unproductive—in a portion of the field? Are there buffers that protect streams, drainage ditches, wellheads, and other water collection points? How far away are the neighbors? What's the general wind direction? This is the place to note these and other special conditions that need to be considered.
- *Recommended rates:* Here's the place where science, technology, and art meet. Given everything you've noted, what is the optimum rate of N, P, K, lime and any other nutrients? While science tells us that a crop has changing nutrient requirements during the growing season, a combination of technology and farmer's management skills assure optimum nutrient availability at all stages of growth. No-till corn generally requires starter fertilizer to give the seedling a healthy start.
- *Recommended timing:* When does the soil temperature drop below 50 degrees? Will a N stabilizer be used? What's the tillage practice? Strip-till corn and no-till often require different timing approaches than seed planted into a field that's been tilled once with a field cultivator. Will a starter fertilizer be used to give the seedling a healthy start? How many acres can be covered with available labor (custom or hired) and equipment? Does manure application in a farm depend on a custom applicator's schedule? What agreements have been worked out with neighbors for manure use on their fields? Is a neighbor hosting a special event? All these factors and more will likely figure into the recommended timing.
- *Recommended methods:* Surface or injected? While injection is clearly preferred, there may be situations where injection is not feasible (i.e. pasture, grassland). Slope, rainfall patterns, soil type, crop rotation and many other factors determine which method is best for optimizing nutrient efficiency (availability and loss) in farms. The combination that's right in one field may differ in another field even with the same crop.
- *Annual review and update:* Even the best managers are forced to deviate from their plans. What rate was actually applied? Where? Using which method? Did an unusually mild winter or wet spring reduce soil nitrate? Did a dry summer, disease, or some other unusual factor increase nutrient carryover? These and other factors should be noted. It's easier to make notes throughout the year than to remember back six to 10 months.

When such a plan is designed for animal feeding operations (AFO), it may be termed a "manure management plan." In the United States, some regulatory agencies recommend or require that farms implement these plans in order to prevent water pollution. The U.S. Natural Resources Conservation Service (NRCS) has published guidance documents on preparing a comprehensive nutrient management plan (CNMP) for AFOs.

The International Plant Nutrition Institute has published a 4R plant nutrition manual for improving the management of plant nutrition. The manual outlines the scientific principles behind each of the four R's or "rights" and discusses the adoption of 4R practices on the farm, approaches to nutrient management planning, and measurement of sustainability performance.

Contour Plowing

Contour ploughing or contour farming or Contour bunding is the farming practice of plowing and/or planting across a slope following its elevation contour lines. These contour lines create a water break which reduces the formation of rills and gullies during times of heavy water run-off; which is a major cause of soil erosion. The water break also allows more time for the water to settle into the soil. In contour ploughing, the ruts made by the plow run perpendicular rather than parallel to slopes, generally resulting in furrows that curve around the land and are level. This method is also known for preventing tillage erosion. Tillage erosion is the soil movement and erosion by tilling a given plot of land. A similar practice is contour bunding where stones are placed around the contours of slopes.

Contour plowing, Pennsylvania, 1938

Soil erosion prevention practices such as this can drastically decrease negative affects associated with soil erosion such as reduced crop productivity, worsened water quality, lower effective reservoir water levels, flooding, and habitat destruction. Contour farming is considered an active form of sustainable agriculture.

"Contour bunding", Catalonia, 2007

History

The Phoenicians first developed the practice of contour farming and spread it throughout the Mediterranean. However, the Romans preferred cultivation in straight furrows and this practice became standard.

In recent years, contour farming systems are considered as field and farm practices to build climate-smart agriculture as they improve water-use efficiency and conservation.

Modern History

This was one of the main procedures promoted by the US Soil Conservation Service (the current Natural Resources Conservation Service) during the 1930s. The US Department of Agriculture established the Soil Conservation Service in 1935 during the Dust Bowl when it became apparent that soil erosion was a huge problem along with desertification.

The extent of the problem was such that the 1934 "Yearbook of Agriculture" noted that *Approximately 35 million acres [142,000 km^2] of formerly cultivated land have essentially been destroyed for crop production. . . . 100 million acres [405,000 km^2] now in crops have lost all or most of the topsoil; 125 million acres [506,000 km^2] of land now in crops are rapidly losing topsoil.* This can lead to large scale desertification which can permanently transform a formerly productive landscape to an arid one that becomes increasingly intensive and expensive to farm.

The Soil Conservation Service worked with state governments and universities with established agriculture programs such as the University of Nebraska to promote the method to farmers. By 1938, the introduction of new agricultural techniques such as contour plowing had reduced the loss of soil by 65% despite the continuation of the drought.

Demonstrations showed that contour farming, under ideal conditions, will increase yields of row crops by up to 50%, with increases of between 5 and 10% being common. Importantly, the technique also significantly reduces soil erosion, fertilizer loss, and overall makes farming less energy and resource intensive under most circumstances.

Reducing fertilizer loss not only saves the farmer time and money, but it also decreases risk of harming regional freshwater systems. Soil erosion caused from heavy rain can encourage the development of rills and gullies which carry excess nutrients into freshwater systems through the process of eutrophication

Contour plowing is also promoted in countries with similar rainfall patterns to the United States such as western Canada and Australia.

The practice is effective only on slopes with between 2% and 10% gradient and when rainfall does not exceed a certain amount within a certain period. On steeper slopes and areas with greater rainfall, a procedure known as strip cropping is used with contour farming to provide additional protection. Contour farming is most effective when used with other soil conservation methods like strip cropping, terrace (agriculture) farming, and the use of a cover crop. The proper combination of such farming methods can be determined by various climatic and soil conditions of that given area. Farming sites are often classified into five levels: insensitive, mild, moderate, high and extreme, depending on the regions soil sensitivity.

P. A. Yeomans' Keyline Design system is critical of traditional contour plowing techniques, and improves the system through observing normal land form and topography. At one end of a contour the slope of the land will always be steeper than at the other. Thus when plowing parallel runs paralleling any contour the plow furrows soon deviate from a true contour. Rain water in these furrows will thus flow sideways along the falling "contour" line. This can often concentrate water in a ways that exacerbates erosion instead of reducing it. Yeomans was the first to appreciate the significance of this phenomenon. Keyline cultivation utilizes this "off contour" drift in cultivating furrows to control the movement of rain water for the benefit of the land. (See Chapter 7 in Priority One History of Twentieth Century Soil Conservation and Keyline.)

Contour bunding has been widely adopted in Burkina Faso after it was suggested by British Oxfam worker Bill Hereford in the beginning of the 1980s.

Its simple definition is that it is a process of parallel ploughing to the contours of hill slopes to form a natural barrier for water to flow down the slope.

Soil Salinity Control

SegReg program: yield of mustard (colza) and soil salinity

Soil salinity control relates to controlling the problem of soil salinity and reclaiming salinized agricultural land.

The aim of soil salinity control is to prevent soil degradation by salination and reclaim already salty (saline) soils. Soil reclamation is also called soil improvement, rehabilitation, remediation, recuperation, or amelioration.

The primary man-made cause of salinization is irrigation. River water or groundwater used in irrigation contains salts, which remain behind in the soil after the water has evaporated.

The primary method of controlling soil salinity is to permit 10-20% of the irrigation water to leach the soil, be drained and discharged through an appropriate drainage system. The salt concentration of the drainage water is normally 5 to 10 times higher than that of the irrigation water, thus salt export matches salt import and it will not accumulate.

The Soil Salinity Problem

Salty (saline) soils are soils that have a high salt content. The predominant salt is normally sodium chloride (NaCl, "table salt"). Saline soils are therefore also *sodic soils* but there may be sodic soils that are not saline, but alkaline.

World Soil Salt Degradation

This damage is an average of 2,000 hectares of irrigated land in arid and semi-arid areas daily for more than 20 years across 75 countries (each week the world loses an area larger than Manhattan)...To feed the world's anticipated nine billion people by 2050, and with little new productive land available, it's a case of all lands needed on deck.—*principal author Manzoor Qadir, Assistant Director, Water and Human Development, at UN University's Canadian-based Institute for Water, Environment and Health*

According to a study by UN University, about 62 million hectares (20%) of the world's irrigated lands are affected, up from 45 million hectares in the early 1990s. In the Indo-Gangetic Plain, home to over 10% of the world's population, crop yield losses for wheat, rice, sugarcane and cotton grown on salt-affected lands could be 40%, 45%, 48%, and 63%, respectively.

Salty soils are a common feature and an environmental problem in irrigated lands in arid and semi-arid regions, resulting in poor or little crop production. The problems are often associated with high water tables, caused by a lack of natural subsurface drainage to the underground. Poor subsurface drainage may be caused by insufficient transport capacity of the aquifer or because water cannot exit the aquifer, for instance if the aquifer is situated in a topographical depression.

Worldwide, the major factor in the development of saline soils is a lack of precipitation. Most naturally saline soils are found in (semi)arid regions and climates of the earth.

Primary Cause

For example, assuming irrigation water with a low salt concentration of 0.3 g/l (equal to

0.3 kg/m³ corresponding to an electric conductivity of about 0.5 FdS/m) and a modest annual supply of irrigation water of 10,000 m³/ha (almost 3 mm/day) brings 3,000 kg salt/ha each year. In the absence of sufficient natural drainage (as in waterlogged soils) and without a proper leaching and drainage program to remove salts, this would lead to a high soil salinity and reduced crop yields in the long run.

Irrigated saline land with poor crop stand

Much of the water used in irrigation has a higher salt content than in this example, which is compounded by the fact that many irrigation projects use a far greater annual supply of water. Sugar cane, for example, needs about 20,000 m³/ha of water per year. As a result, irrigated areas often receive more than 3,000 kg/ha of salt per year and some receive as much as 10,000 kg/ha/year.

Secondary Cause

The secondary cause of salinization is waterlogging in irrigated land. Irrigation causes changes to the natural water balance of irrigated lands. Large quantities of water in irrigation projects are not consumed by plants and must go somewhere. In irrigation projects it is impossible to achieve 100% irrigation efficiency where all the irrigation water is consumed by the plants. The maximum attainable irrigation efficiency is about 70% but usually it is less than 60%. This means that minimum 30%, but usually more than 40% of the irrigation water is not evaporated and it must go somewhere.

Most of the water lost this way is stored underground which can change the original hydrology of local aquifers considerably. Many aquifers cannot absorb and transport these quantities of water and so the water table rises leading to water logging.

Waterlogging causes three problems:

- The shallow water table and lack of oxygenation of the root zone reduces the yield of most crops
- It leads to an accumulation of salts brought in with the irrigation water as their removal through the aquifer is blocked

- With the upward seepage of groundwater more salts are brought into the soil and the salination is aggravated

Aquifer conditions in irrigated land and the groundwater flow have an important role in soil salinization, as illustrated here :

- Illustration of the influence of aquifer conditions on soil salinization in irrigated land

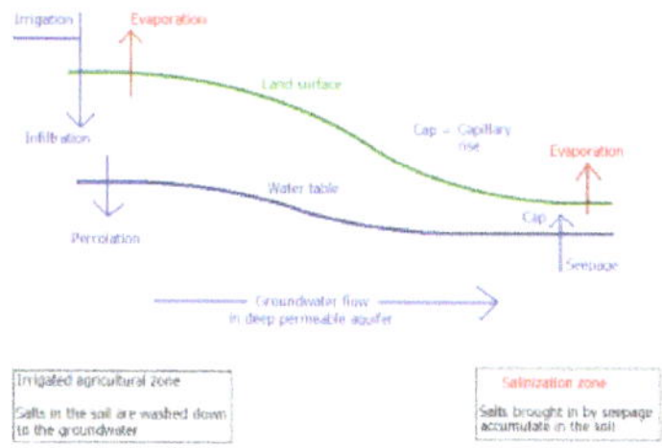

Soil salinization in the lower parts of undulating land with a good aquifer

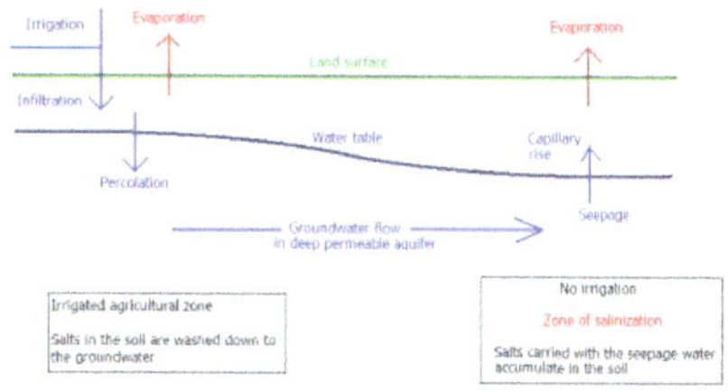

Soil salinization in the unirrigated parts of flat land with a good aquifer

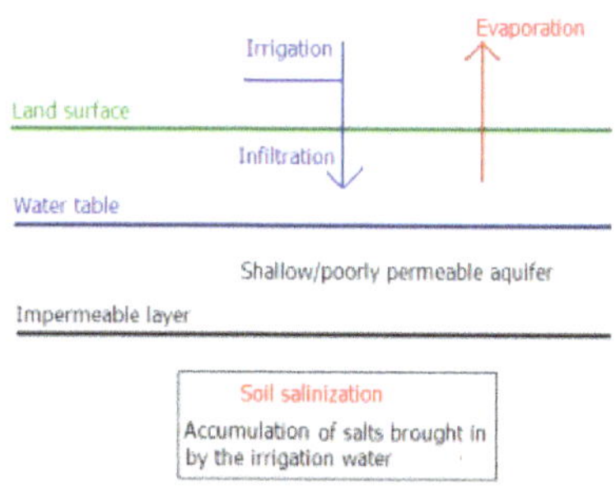

Soil salinization in irrigated flat land without an aquifer

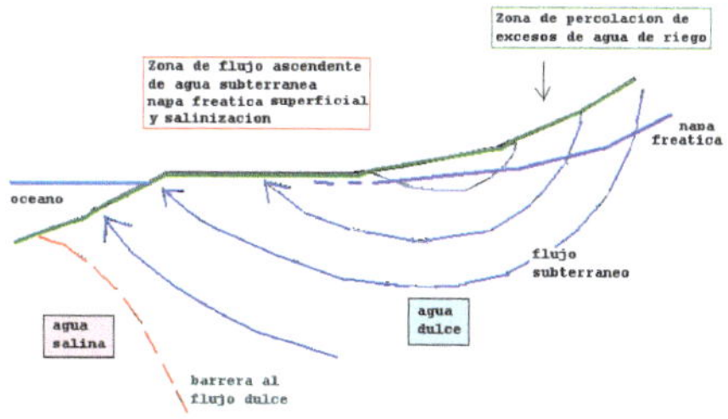

Soil salinization in a coastal delta from irrigation higher up

Salt Affected Area

Normally, the salinization of agricultural land affects a considerable area of irrigation projects, on the order of 20 to 30%. When the agriculture in such a fraction of the land is abandoned, a new salt and water balance is attained, a new equilibrium is reached, and the situation becomes stable.

In India alone, thousands of square kilometres have been severely salinized. China and Pakistan do not lag much behind (perhaps China has even more salt affected land than India). A regional distribution of the 3,230,000 km^2 of saline land worldwide is shown in the following table derived from the FAO/UNESCO Soil Map of the World.

Region	Area (10^6ha)
Australia	84.7
Africa	69.5
Latin America	59.4
Near and Middle East	53.1
Europe	20.7
Asia and Far East	19.5
Northern America	16.0

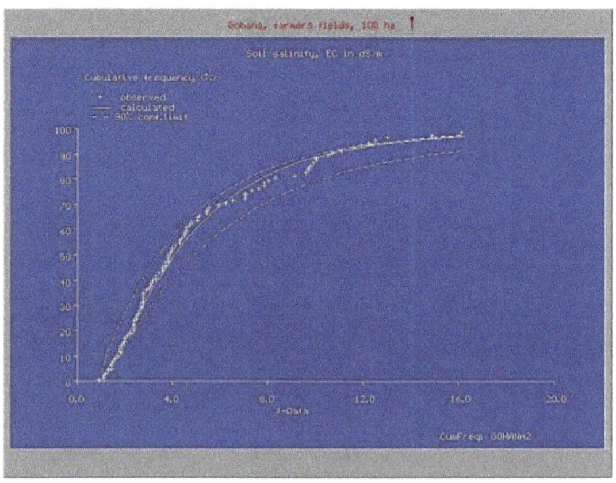

CumFreq program: spatial variation of soil salinity

Spatial Variation

Although the principles of the processes of salinization are fairly easy to understand, it is more difficult to explain why certain parts of the land suffer from the problems and other parts do not, or to predict accurately which part of the land will fall victim. The main reason for this is the variation of natural conditions in time and space, the usually uneven distribution of the irrigation water, and the seasonal or yearly changes of agricultural practices. Only in lands with undulating topography is the prediction simple: the depressional areas will degrade the most.

The preparation of salt and water balances for distinguishable sub-areas in the irrigation project, or the use of agro-hydro-salinity models, can be helpful in explaining or predicting the extent and severity of the problems.

Diagnosis

Measurement

Soil salinity is measured as the salt concentration of the soil solution in tems of g/l or electric conductivity (EC) in dS/m. The relation between these two units is about 5/3 : y g/l => 5y/3 dS/m. Seawater may have a salt concentration of 30 g/l (3%) and an EC of 50 dS/m.

The standard for the determination of soil salinity is from an extract of a saturated paste of the soil, and the EC is then written as ECe. The extract is obtained by centrifugation. The salinity can more easily be measured, without centrifugation, in a 2:1 or 5:1 water:soil mixture (in terms of g water per g dry soil) than from a saturated paste. The relation between ECe and $EC_{2:1}$ is about 4, hence : ECe = 4 $EC_{1:2}$.

Classification

Soils are considered saline when the ECe > 4. When 4 < ECe < 8, the soil is called slightly saline, when 8 < ECe < 16 it is called (moderately) saline, and when ECe > 16 severely saline.

Crop Tolerance

Sensitive crops lose their vigor already in slightly saline soils, most crops are negatively affected by (moderately) saline soils, and only salinity resistant crops thrive in severely saline soils. The University of Wyoming and the Government of Alberta report data on the salt tolerance of plants.

Principles of Salinity Control

Drainage is the primary method of controlling soil salinity. The system should permit a small fraction of the irrigation water (about 10 to 20 percent, the drainage or leaching fraction) to be drained and discharged out of the irrigation project.

In irrigated areas where salinity is stable, the salt concentration of the drainage water is normally 5 to 10 times higher than that of the irrigation water. Salt export matches salt import and salt will not accumulate.

When reclaiming already salinized soils, the salt concentration of the drainage water will initially be much higher than that of the irrigation water (for example 50 times higher). Salt export will greatly exceed salt import, so that with the same drainage fraction a rapid desalinization occurs. After one or two years, the soil salinity is decreased so much, that the salinity of the drainage water has come down to a normal value and a new, favorable, equilibrium is reached.

In regions with pronounced dry and wet seasons, the drainage system may be operated in the wet season only, and closed during the dry season. This practice of checked or controlled drainage saves irrigation water.

The discharge of salty drainage water may pose *environmental problems* to downstream areas. The environmental hazards must be considered very carefully and, if necessary mitigating measures must be taken. If possible, the drainage must be limited to wet seasons only, when the salty effluent inflicts the least harm.

Drainage Systems

The drainage system designed to evacuate salty water also lowers the water table. To reduce the cost of the system, the lowering must be reduced to a minimum. The highest permissible level of the water table (or the shallowest permissible depth) depends on the irrigation and agricultural practices and kind of crops.

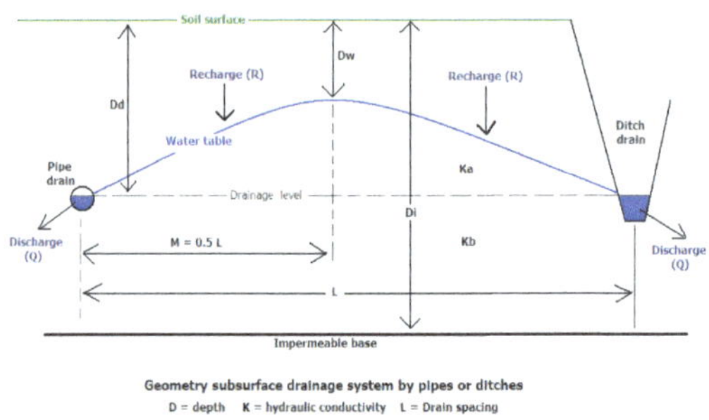

Parameters of a horizontal drainage system

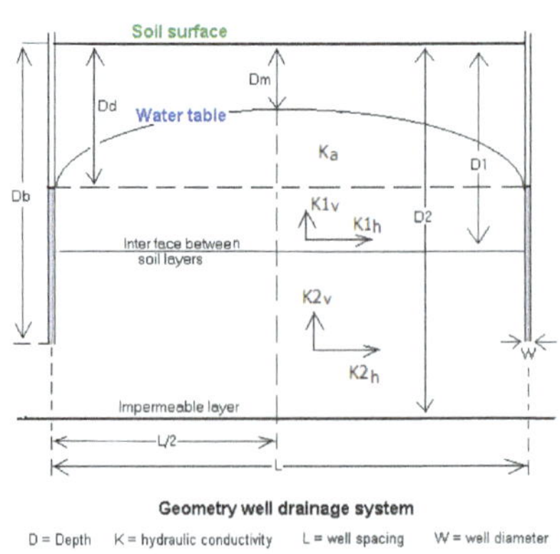

Parameters of a vertical drainage system

Land drainage for soil salinity control is usually by *horizontal drainage system* (figure left), but *vertical systems* (figure right) are also employed.

In many cases a seasonal average water table depth of 0.6 to 0.8 m is deep enough. This means that the water table may occasionally be less than 0.6 m (say 0.2 m just after an irrigation or a rain storm). This automatically implies that, in other occasions, the water table will be deeper than 0.8 m (say 1.2 m). The fluctuation of the water table helps in the breathing function of the soil while the expulsion of carbon dioxide (CO_2) produced by the plant roots and the inhalation of fresh oxygen (O_2) is promoted.

The establishing of a not too deep water table offers the additional advantage that excessive field irrigation is discouraged, as the crop yield would be negatively affected by the resulting elevated water table, and irrigation water may be saved.

The statements made above on the optimum depth of the water table are very general, because in some instances the required water table may be still shallower than indicated (for example in rice paddies), while in other instances it must be considerably deeper (for example in some orchards). The establishment of the optimum depth of the water table is in the realm of agricultural drainage criteria.

Soil Leaching

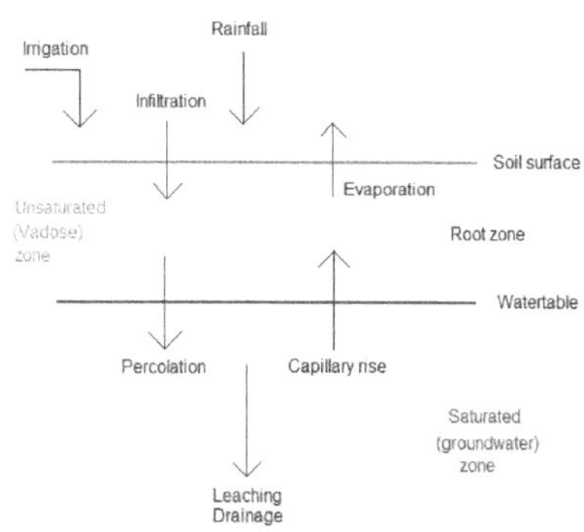

Water balance factors in the soil

The vadose zone of the soil below the soil surface and the watertable is subject to four main hydrological inflow and outflow factors:

- Infiltration of rain and irrigation water (Irr) into the soil through the soil surface (Inf) :
- Inf = Rain + Irr
- Evaporation of soil water through plants and directly into the air through the soil surface (Evap)
- Percolation of water from the unsaturated zone soil into the groundwater through the watertable (Perc)
- Capillary rise of groundwater moving by capillary suction forces into the unsaturated zone(Cap)

In steady state (i.e. the amount of water stored in the unsaturated zone does not change in the long run) the water balance of the unsaturated zone reads: Inflow = Outflow, thus:

- Inf + Cap = Evap + Perc or :
- Irr + Rain + Cap = Evap + Perc

and the *salt balance* is

- Irr.Ci + Cap.Cc = Evap.Fc.Ce + Perc.Cp + Ss

where Ci is the salt concentration of the irrigation water, Cc is the salt concentration of the capillary rise, equal to the salt concentration of the upper part of the ground-water body, Fc is the fraction of the total evaporation transpired by plants, Ce is the salt concentration of the water taken up by the plant roots, Cp is the salt concentration of the percolation water, and Ss is the increase of salt storage in the unsaturated soil. This assumes that the rainfall contains no salts. Only along the coast this may not be true. Further it is assumed that no runoff or surface drainage occurs. The amount of removed by plants (Evap.Fc.Ce) is usually negligibly small: Evap.Fc.Ce = 0

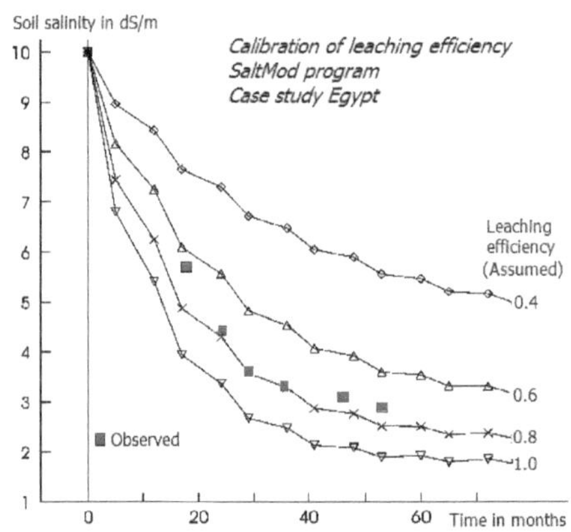

Leaching curves, calibrating leaching efficiency

The salt concentration Cp can be taken as a part of the salt concentration of the soil in the unsaturated zone (Cu) giving: Cp=Le.Cu, where Le is the leaching efficiency. The leaching efficiency is often in the order of 0.7 to 0.8, but in poorly structured, heavy clay soils it may be less. In the Leziria Grande polder in the delta of the Tagus river in Portugal it was found that the leaching efficiency was only 0.15. Assuming that one wishes to avoid the soil salinity to increase and maintain the soil salinity Cu at a desired level Cd we have:Ss = 0, Cu = Cd and Cp = Le.Cd. Hence the salt balance can be simplified to:

- Perc.Le.Cd = Irr.Ci + Cap.Cc

Setting the amount percolation water required to fulfill this salt balance equal to Lr (the *leaching requirement*) it is found that:

- $Lr = (Irr.Ci + Cap.Cc)/Le.Cd$.

Substituting herein $Irr = Evap + Perc - Rain - Cap$ and re-arranging gives :

- $Lr = [\ (Evap-Rain).Ci + Cap(Cc-Ci)\ ]/(Le.Cd - Ci)$

With this the irrigation and drainage requirements for salinity control can be computed too. In irrigation projects in (semi)arid zones and climates it is important to check the leaching requirement, whereby the *field irrigation efficiency* (indicating the fraction of irrigation water percolating to the underground) is to be taken into account. The desired soil salinity level Cd depends on the crop tolerance to salt. The University of Wyoming, USA, and the Government of Alberta, Canada, report crop tolerance data.

Strip Cropping: An Alternative

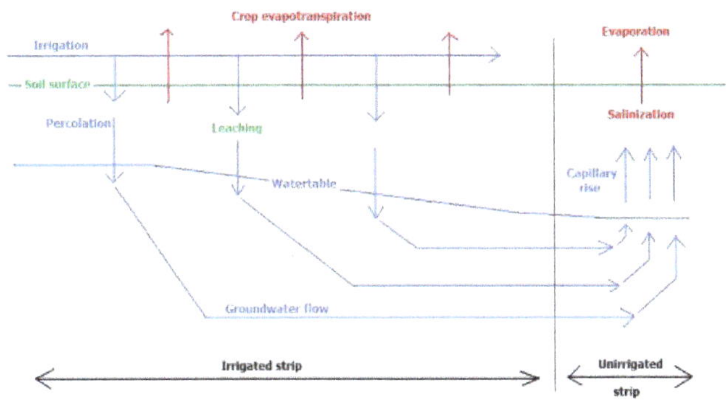

Principle of strip cropping and land sacrifice

Hydrological principles of *strip cropping* to control the depth of the water table and the soil salinity

In irrigated lands with scarce water resources suffering from drainage (high water table) and soil salinity problems, strip cropping is sometimes practiced with strips of land where every other strip is irrigated while the strips in between are left permanently fallow.

Owing to the water application in the irrigated strips they have a higher watertable which induces flow of groundwater to the unirrigated strips. This flow functions as subsurface drainage for the irrigated strips, whereby the water table is maintained at a not-too-shallow depth, leaching of the soil is possible, and the soil salinity can be controlled at an acceptably low level.

In the unirrigated (sacrificial) strips the soil is dry and the groundwater comes up by capillary rise and evaporates leaving the salts behind, so that here the soil salinizes. Nevertheless, they can have some use for livestock, sowing salinity resistant grasses or weeds. Moreover, useful salt resistant trees can be planted like Casuarina, Eucalyptus or Atriplex, keeping in mind that the trees have deep rooting systems and the salinity of the wet subsoil is less than of the topsoil. In these ways wind erosion can be controlled. The unirrigated strips can also be used for salt harvesting.

Soil Salinity Models

The majority of the computer models available for water and solute transport in the soil (e.g. SWAP, DrainMod-S, UnSatChem, and Hydrus) are based on Richard's differential equation for the movement of water in unsaturated soil in combination with Fick's differential convection–diffusion equation for advection and dispersion of salts.

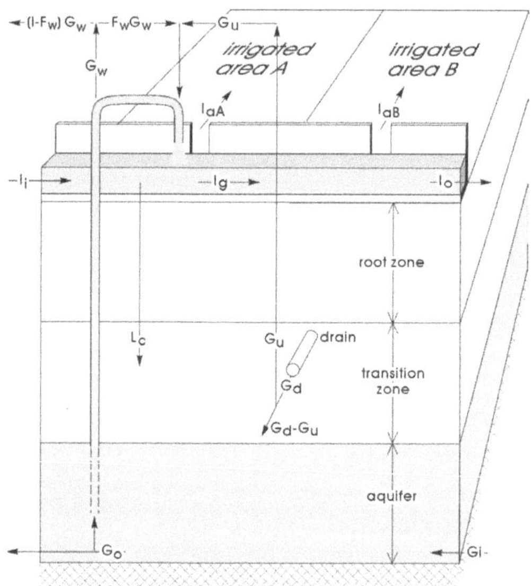

SaltMod components

The models require input of soil characteristics like the relations between variable unsaturated soil moisture content, water tension, water retention curve, unsaturated hydraulic conductivity, dispersivity and diffusivity. These relations vary to a great extent from place to place and from time to time and are not easy to measure. Further, the models are difficult to calibrate under farmer's field conditions because the soil salinity here is spatially very variable. The models use short time steps and need at least a daily, if not an hourly, data base of hydrological phenomena. Altogether this makes model application to a fairly large project the job of a team of specialists with ample facilities.

Simpler models, like SaltMod, based on monthly or seasonal water and soil balances and an empirical capillary rise function, are also available. They are useful for long-term salinity predictions in relation to irrigation and drainage practices.

LeachMod, using the SaltMod principles, helps in analyzing leaching experiments in which the soil salinity was monitored in various root zone layers while the model will optimize the value of the leaching efficiency of each layer so that a fit is obtained of observed with simulated soil salinity values.

Spatial variations owing to variations in topography can be simulated and predicted using salinity cum groundwater models, like SahysMod.

References

- Arthur T. Hubbard, Encyclopedia of Surface and Colloid Science Vol 3, Santa Barbara, California Science Project, Marcel Dekker, New York (2004) ISBN 0-8247-0759-1
- Alfred J. Turgeon; Lambert Blanchard McCarty; Nick Edward Christians (2009). Weed control in turf and ornamentals. Prentice Hall. p. 126. ISBN 978-0-13-159122-6.
- Mahesh K. Upadhyaya; Robert E. Blackshaw (2007). Non-chemical Weed Management: Principles, Concepts and Technology. CABI. pp. 135–. ISBN 978-1-84593-291-6.
- Stout, Ruth. Gardening Without Work. Devon-Adair Press, 1961. Reprinted by Norton Creek Press, 2011, pp. 6-7. ISBN 978-0-9819284-6-3
- Jeff Gillman (1 February 2008). The Truth About Organic Gardening: Benefits, Drawnbacks, and the Bottom Line. Timber Press. pp. 51–. ISBN 978-1-60469-005-7.
- David A. Bainbridge (11 June 2007). A Guide for Desert and Dryland Restoration: New Hope for Arid Lands. Island Press. pp. 239–. ISBN 978-1-61091-082-8
- Walker, John (2011). How to Create an Eco Garden: The practical guide to greener, planet-friendly gardening. Wigston, Leicestershire: Aquamarine. p. 33. ISBN 9781903141892.
- Arthur T. Hubbard, Encyclopedia of Surface and Colloid Science Vol 3, Santa Barbara, California Science Project, Marcel Dekker, New York (2004) ISBN 0-8247-0759-1
- Goode, Erica (March 10, 2015). "Farmers Put Down the Plow for More Productive Soil". The New York Times (New York ed.). The New York Times Company. p. D1. ISSN 0362-4331. OCLC 1645522. Retrieved April 5, 2015.
- Ray Hilborn. "Soils in Agriculture" (PPT--available as non-PPT by searching the path through a search engine). University of Washington. Retrieved 2013-08-28.
- Mahdi Al-Kaisi; Mark Hanna; Michael Tidman (13 May 2002). "Methods for measuring crop residue". Iowa State University. Retrieved 2012-12-28.
- "Soil Compaction and Conservation Tillage". Conservation Tillage Series. PennState- College of Agricultural Sciences - Cooperative Extension. Retrieved 26 March 2011.

Permissions

All chapters in this book are published with permission under the Creative Commons Attribution Share Alike License or equivalent. Every chapter published in this book has been scrutinized by our experts. Their significance has been extensively debated. The topics covered herein carry significant information for a comprehensive understanding. They may even be implemented as practical applications or may be referred to as a beginning point for further studies.

We would like to thank the editorial team for lending their expertise to make the book truly unique. They have played a crucial role in the development of this book. Without their invaluable contributions this book wouldn't have been possible. They have made vital efforts to compile up to date information on the varied aspects of this subject to make this book a valuable addition to the collection of many professionals and students.

This book was conceptualized with the vision of imparting up-to-date and integrated information in this field. To ensure the same, a matchless editorial board was set up. Every individual on the board went through rigorous rounds of assessment to prove their worth. After which they invested a large part of their time researching and compiling the most relevant data for our readers.

The editorial board has been involved in producing this book since its inception. They have spent rigorous hours researching and exploring the diverse topics which have resulted in the successful publishing of this book. They have passed on their knowledge of decades through this book. To expedite this challenging task, the publisher supported the team at every step. A small team of assistant editors was also appointed to further simplify the editing procedure and attain best results for the readers.

Apart from the editorial board, the designing team has also invested a significant amount of their time in understanding the subject and creating the most relevant covers. They scrutinized every image to scout for the most suitable representation of the subject and create an appropriate cover for the book.

The publishing team has been an ardent support to the editorial, designing and production team. Their endless efforts to recruit the best for this project, has resulted in the accomplishment of this book. They are a veteran in the field of academics and their pool of knowledge is as vast as their experience in printing. Their expertise and guidance has proved useful at every step. Their uncompromising quality standards have made this book an exceptional effort. Their encouragement from time to time has been an inspiration for everyone.

The publisher and the editorial board hope that this book will prove to be a valuable piece of knowledge for students, practitioners and scholars across the globe.

Index